aviation & space
education

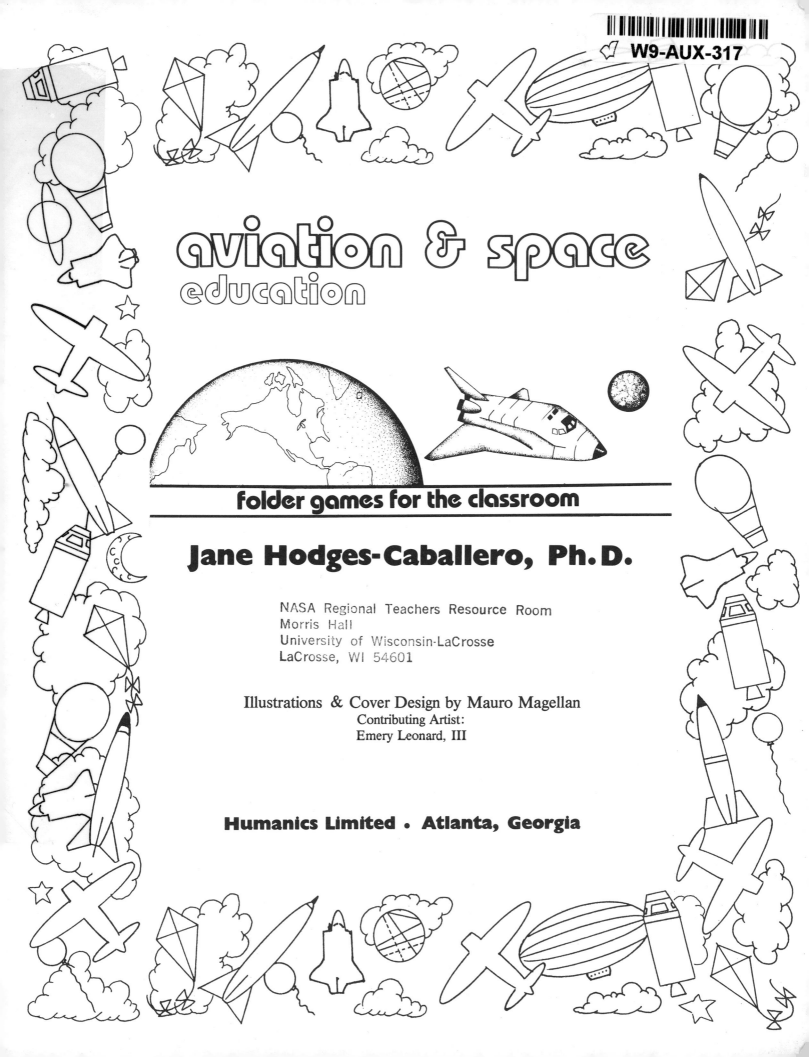

folder games for the classroom

Jane Hodges-Caballero, Ph.D.

Illustrations & Cover Design by Mauro Magellan
Contributing Artist:
Emery Leonard, III

Humanics Limited • Atlanta, Georgia

Also by Jane Hodges-Caballero:

Vanilla Manila Folder Games
The Handbook of Learning Activities for Young Children
Month by Month Activity Guide for the Primary Grades
Art Projects for Young Children
Aerospace Projects for Young Children

Co-authored by Jane Hodges-Caballero:

Back To Basics in Reading Made Fun
Children Around the World
Humanics National Infant & Toddler Assessment Handbook

HUMANICS LIMITED
P.O. Box 7447
Atlanta, Georgia 30309

PRINTED IN THE UNITED STATES OF AMERICA

ISBN: 0 - 89334 - 087 - 1

LIBRARY OF CONGRESS NO: 85-081658

Illustrations & Cover Design by Mauro Magellan
Contributing Artist:
Emery Leonard, III

Humanics Limited • Atlanta, Georgia

Contents

IV. Resources

V. Aerospace For the Classroom

Preface

Aviation and Space Education will provide a broad foundation on which to build a lifelong interest in all things using our atmosphere and the vast space above it as a medium (or non-medium) through which to travel.

Activities in this book are designed to stimulate excitement and interest in aviation and space. Each activity is keyed to a basic skill such as reading, writing, mathematics, science, and social studies. Activities not only provide skill building in various disciplines; each activity is also designed to enhance the student's basic understanding of aviation and rocketry.

I encourage you to give your students a head start into the 21st Century.

Ted Colton
Aerospace Education Workshop Director
Professor of Science Education
Georgia State University
Atlanta, Georgia

Acknowledgements

I would like to extend my sincere appreciation to the following persons for their encouragment and continued support in promoting aerospace education:

The Federal Aviation Administration; Congressman Don Clausen, Director of Special Programs and Jack K. Barker, Public Affairs/Aviation Education Officer, Southern Region.

The U.S.A.F. Civil Air Patrol: Harold (Hal) R. Bacon, Kenneth Perkins, Colonel James R. Withers, and Phillip S. Woodruff

University Aviation Association: Gary Kiteley, Executive Director

World Aerospace Education Organization: Dr. Mervin K. Strickler, Jr, Jack Sorenson and Kamal Naguib

The National Aeronautics and Space Administration: Dr. Curtis M. Graves, Office of Educational Affairs, Washington, D.C.

Thanks also go to Dr. Anne Christenberry, Augusta College, Augusta, Georgia; Drs. Ben DeMayo and Glenda Norwood of West Georgia College; and Ms. Carol Hickson, of The Fernbank Science Center in Atlanta, Georgia for encouraging field testing of these activities in the teacher Aerospace Education Workshops.

Introduction

A simple file folder, these patterns, and colored marking pens are all you need to make your own Aviation and Space Folder Games. You can make variations as suggested or use your own imagination to individualize folders based on the student's developmental level.

These unique folders encourage the educator to take an active role in designing curriculum for the students. It is not a ready-made kit that merely requires presentation.

Students can readily identify with the folders and are enthusiastic when they see their teacher actually making them. They are also motivated when allowed to make their own folders.

These practical folders not only are inexpensive to develop, but also are easy to store and use in a learning center. The curriculum folders meet the educational objectives in all subject areas for upper elementary and middle grade students. However, young students may profit from the folders since many do not have an aviation/space background. They require only a minimum amount of adult instruction. Therefore, they are ideal for center time and parent use as well. They begin with boomerangs and kites and end with the Space Shuttle and Space Stations.

Opportunities for using the folder games are endless. Boredom and frustration can be eliminated and the student's learning process can be stimulated.

Be sure to use bright, colorful watercolor markers. Color the cover patterns, then glue the cover and inside patterns to your file folder. Next, cut out the directions and glue to the top of the back cover. Some folders may have additional activities on the bottom of the back cover. The miniature photographs can be glued onto index cards and kept by the teacher for a check-out file, index of activities or other management techniques. Laminate and have fun learning!

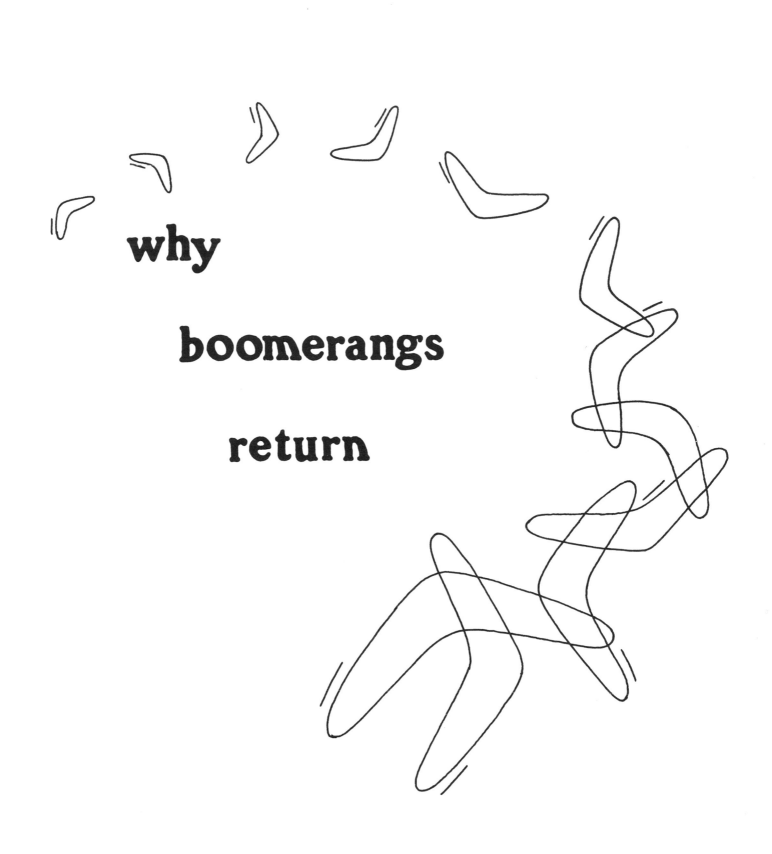

why

boomerangs

return

NAME: Boomerangs

SKILL: Motor Skill Development

PROCEDURE: Read about boomerangs, then make a small boomerang to throw out of popsicle sticks, tongue depressors or cardboard. See how accurately and how far you can throw a boomerang.

VARIATIONS: Play the game with your classmates.

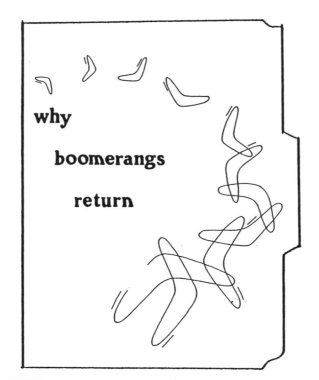

why

boomerangs

return

Boomerangs were found more than 10,000 years ago in Australia. Boomerangs were used for killing, sport, ceremonies and other purposes. Why does a boomerang return? Simply stated 1) The wing of the boomerang has an airfoil and generates lift. 2) Spinning it produces stability in flight. 3) The spin and forward motion imparted by the thrower causes a gyroscopic action to occur and the boomerang turns and turns and returns. The boomerang can travel 60 miles per hour and spins at 10 revolutions per second.

MATERIALS: 2 popsicle sticks or 2 tongue depressors
thread, rubber band or glue
sandpaper

1. Round leading edge with sandpaper and paper back stick to the trailing edge on top side of stick only.
2. Fasten together with rubber band, thread, or glue.
3. Grasp one vane with fingers and toss with an overhand motion.

MATERIALS: index card or piece of cardboard
scissors and pencil

1. Draw the outline of the small boomerang on a piece of cardboard.
2. Cut out the boomerang.
3. Balance the boomerang on your left finger and flick one of the legs sharply with your other finger.

Reprinted with permission from Dr. Ted Colton, Aerospace Resource Center, Georgia State University.

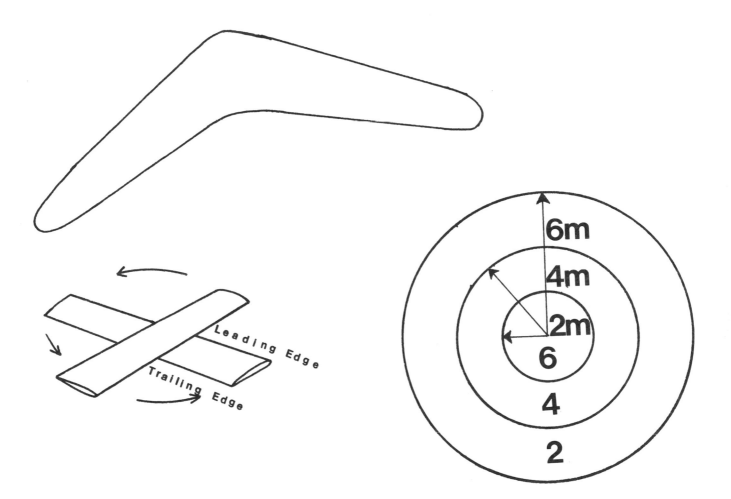

RULES:

1. Contestant throws from center circle. If boomerang lands within a circle, the contestant receives 2, 4 or 6 points.
2. If the contestant catches the boomerang anywhere (even outside the circle) he recieves a bonus of 4 points.
3. Each contestant gets 10 throws, perfect score is 100.
4. If contestant's feet are in different zones when a catch is made, he receives the average of the scores for the circles plus 4 bonus points.
5. If contestant touches the boomerang but fails to catch it, he receives 1 bonus point.

Based on standard rules of Mudgeerabe Creek Emu Racing and Boomerang Throwing Association, Mudgeerabe, Queensland, Australia.

Kites

4

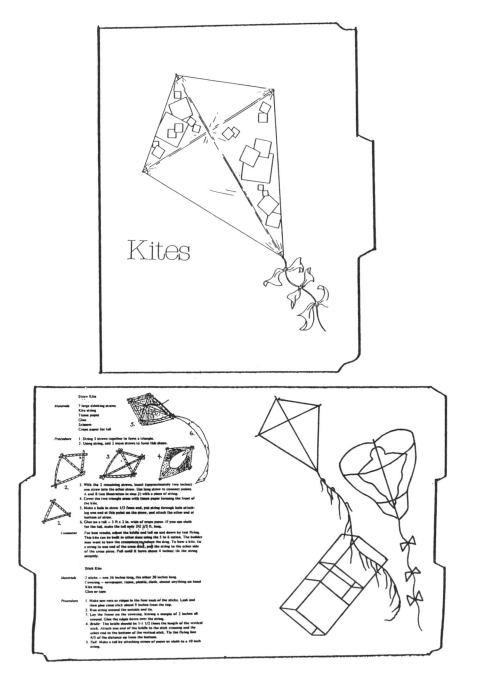

NAME: Kites

SKILL: Science and Art

PROCEDURE: Make a kite from either of the two descriptions. Fly your kite. Compare the flying capabilities of each kite.

VARIATIONS: 1. Encourage the students to make their own designs and kites, then display small models of the kites on the bulletin board.
2. Using library resources, study the history and different types of kites used by the Chinese, Japanese, Europeans and Americans.

Kites

STRAW KITE

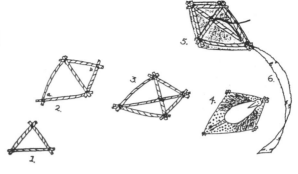

Materials   7 large drinking straws
Kite string
Tissue paper
Glue
Scissors
Crepe paper for tail

Procedure

1. String 3 straws together to form a triangle.
2. Using string, add 2 more straws to form this shape.
3. With the 2 remaining straws, insert (approximately two inches) one straw into the other straw. Use long straw to connect points A and B (see illustration in step 2) with a piece of string.
4. Cover the two triangle areas with tissue paper forming the front of the kite.
5. Make a hole in straw ⅓ from end, put string through hole attaching one end at this point on the straw, then attach the other end at bottom of straw.
6. Glue on tail — 3 ft. X 2 in. wide of crepe paper. If you use cloth for the tail, make the tail only 2½ ft. long.

Comment   For best results, adjust the bridle and tail up and down by test flying. This kite can be built in other sizes using the 5 to 6 ratio. The builder may want to bow the crosspiece to reduce the drag. To bow a kite, tie a string to one end of the cross stick, pull the string to the other side of the cross piece. Pull until it bows about 4 inches; tie the string securely.

STICK KITE

Materials   2 sticks — one 36 inches long, the other 30 inches long.
Covering — newspaper, tissue, plastic, cloth, almost anything on hand
Kite string
Glue or tape

Procedure

1. Make saw cuts or ridges in the four ends of the sticks. Lash and then glue cross stick about 9 inches from the top.
2. Run string around the outside and tie.
3. Lay the frame on the covering, leaving a margin of 2 inches all around. Glue the edges down over the string.
4. *Bridle*: The bridle should be 1½ times the length of the vertical stick. Attach one of the bridle to the stick crossing and the other end to the bottom of the vertical stick. Tie the flying line 4/5 of the distance up from the bottom.
5. *Tail*: Make a tail by attaching scraps of paper or cloth to a 10 inch string.

American Veterans Traveling Tribute

Vietnam Veterans Memorial

1955 - 1975

· LOWELL T GLOSSUP ·

PANEL __34__ LINE __48__

THIS MEMORIAL IS DEDICATED TO THE MEMORY OF THOSE
WHO LOST THEIR LIVES IN THE VIETNAM WAR AND TO THE MISSING IN ACTION
AND PRISONERS OF WAR WHO WERE LEFT BEHIND IN SOUTHEAST ASIA.
"THEY WILL NEVER BE FORGOTTEN"

BALLOONS

NAME: Balloons

SKILL: Reading Comprehension and Art Activity

PROCEDURE: Read the story about balloons. Color the worksheets, noting the parts of the balloon and the various patterns. Be colorful in designing repetitious color patterns. Use cotton to form clouds on the cover.

VARIATIONS: 1. Cover a blown up balloon with papier mache. Paint and decorate your balloon. Attach a basket at the bottom.
2. Create your own balloon patterns and color design.

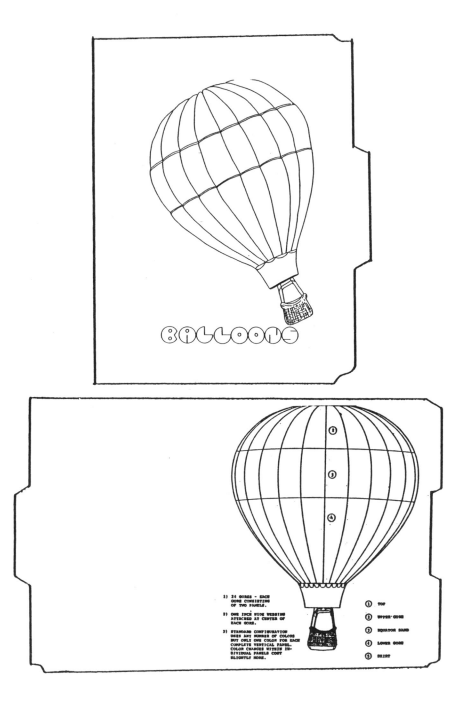

Montgolfier Balloon

BALLOONS

A balloon can rise because a gas is used inside the bag which is much lighter than the air around it. The main gases which are used are hydrogen, helium, and coal gas. Hydrogen is the lightest of the gases, so has the ability to give the balloon the most lift. However, it is dangerous as it is very flammable. Helium is much safer to use, since it does not burn. Coal gas does not lift as well, but it is less expensive.

Riding a balloon is a sport for many poeple who enjoy the thrills and adventures of floating in the air. There is much skill required in locating various currents of air and maintaining the proper altitude to fly with certain paths of air.

There are many national and international sporting events and world record attempts. The F.A.A. has three types of ratings: hot air with airborn burner, hot air without an airborn burner, or free helium or gas balloon. One can learn to fly and obtain a license in a balloon after approximately 10 hours of instruction and one solo flight. A written exam is also required.

The earliest balloons were filled with heated air. Since hot air is less dense than cold air, the balloon would rise. However, the air would soon cool so the ascension of hot air balloons is short.

The material first used for balloon bags were silk or cotton, made airtight with a coating of rubber. Today, plastic nylon and polyster fabrics are used. A balloon may cost from $5,000 to $20,000. Today both heater and valve controls are available.

The Montgolfier brothers experimented with the first hot air balloon and on June 4, 1783; they released a fine, paper-lined balloon which rose to 6,000 feet and traveled more than a mile. Three months later they were invited to Versailles to put on a demonstration for King Louis XVI and Marie Antoinette. They attached a cage to the balloon which contained a sheep, a rooster and a duck. The animals stayed aloft eight minutes and traveled a mile and a half. From this time on all balloons became known as Montgolfiers.

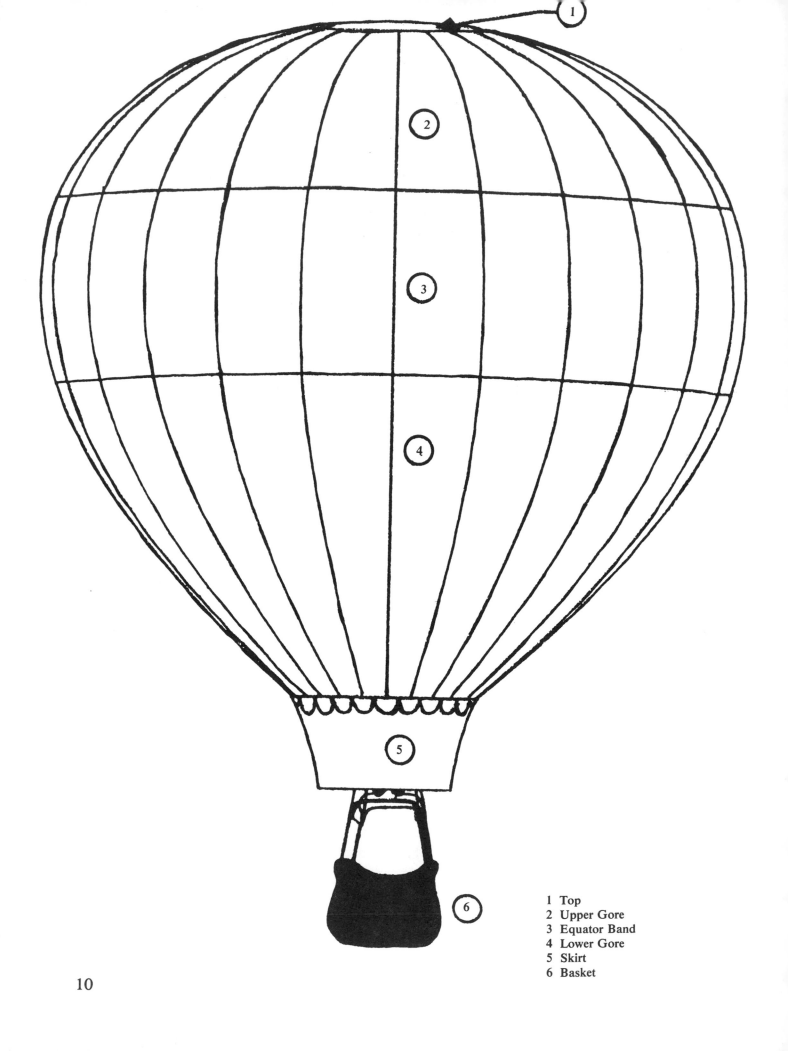

1 Top
2 Upper Gore
3 Equator Band
4 Lower Gore
5 Skirt
6 Basket

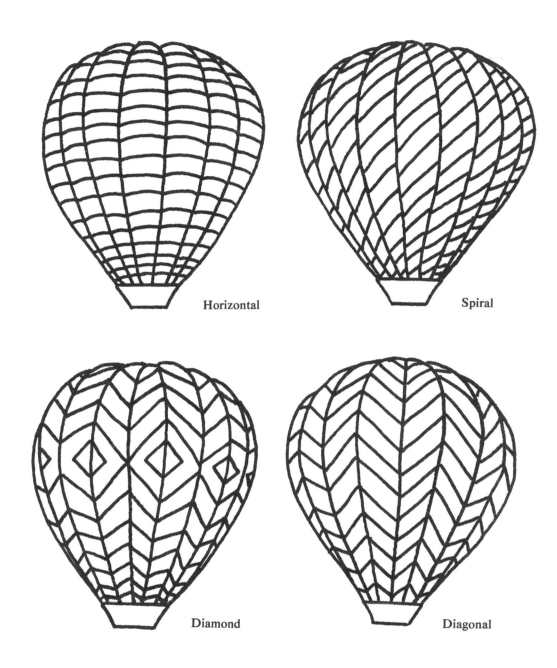

Horizontal

Spiral

Diamond

Diagonal

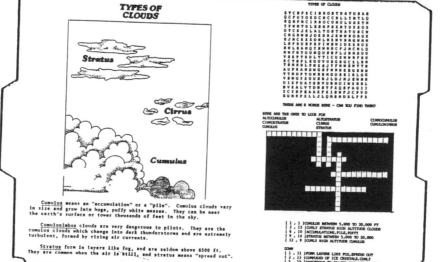

NAME: Cloud Matching

SKILL: Language Activity

PROCEDURE: Read about the different types of clouds. Find the names of the different clouds in the word search game. Complete the crossword puzzle. Check your answers with the answer key on the back of the folder.

VARIATIONS:
1. Let a teakettle boil and watch the escaping vapor condense as it rises and cools, forming a small cloud.
2. Watch the weather reports and see if you can identify the cloud formations.
3. Keep a record of your cloud observations, what you predict and the actual weather.

13

CLOUDS

Clouds are visible vaporous formations in the sky. There are four different types of clouds. Each has its own distinctive characteristics.

Cumulus means "accumulation" or "pile". Cumulus clouds vary in size and grow into huge, puffy white masses. They can be near the earth's surface or tower thousands of feet into the sky.

Cumulonimbus clouds are very dangerous to pilots. They are cumulus clouds which change into dark thunderstorms and are extremely turbulent, formed by rising air currents.

Stratus form in layers like fog, and are seldom above 6500 ft. They are common when the air is still, and stratus means, "spread out".

Cirrus clouds are in the upper levels of the troposphere, usually between 20,000 and 50,000 feet, and are composed of ice crystals which have a delicate, curly appearance like feathers.

"Alto" is generally added to designate clouds at intermediate height, usually at levels between 5,000 and 20,000 feet. These could be *altocumulus* and *altostratus*.

"Cirro" means curly and is added to the high altitude clouds. *Cirrocumulus* and *cirrostratus* are in this group.

TYPES OF CLOUDS

```
E T C B F Z C I R R O S T R A T U S A H
Q C F U Y O K D C H C C R L L Y N T L Q
H Q B P M C I R R O C U M U L U S K T M
Y P W M T U L E S H P S X R Y U X S O X
O T C K J K L A L T O S T R A T U S C W
C M G R W N J O K D R J K W W R O B U K
E J M C K X X O N L M Z N V R M G W M M
L S F R U E P E P I W H W I D K Z R U X
B W L E A M Q Y S W M R C F J M C M L Q
W W F H T U U N Q U S B F L J H M S U U
V S K P T K U L T T I C U X W W R F S U
K C T N P L E G U V U S G S G C Z L B H
T H M R H L P W Z S J Y F M W E W T T H
A U P K A N D W G S F N R H K E G E N N
T T M D P T O W E B M G O E E I R L G G
V S W G Y X U U S B J L D J M J E W N H
U I X F D A Y S H V U B M O I W Z H R L
B V K A L V T D H S T F X Q S F A G I I
Y C Y R P E A A X F X I S W F J C P V U
E U M E P X L L J L Q M S M D X L F F Z
```

THERE ARE 8 WORDS HERE - CAN YOU FIND THEM?

HERE ARE THE ONES TO LOOK FOR

ALTOCUMULUS	ALTOSTRATUS	CIRROCUMULUS
CIRROSTRATUS	CIRRUS	CUMULONIMBUS
CUMULUS	STRATUS	

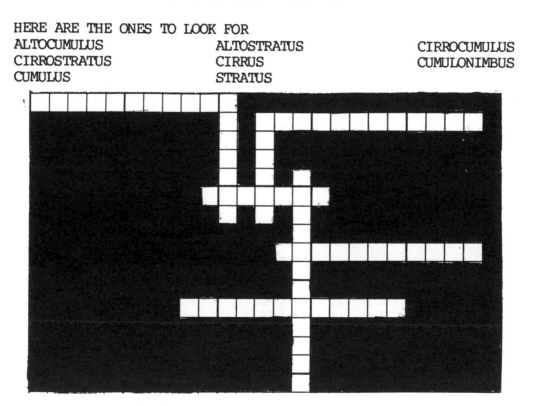

ACROSS

[1 , 1]CUMULUS BETWEEN 5,000 TO 20,000 FT
[2 , 13]CURLY STRATUS HIGH ALTITUDE CLOUDS
[6 , 10]ACCUMULATIONS,PILE,PUFFY
[9 , 14]STRATUS BETWEEN 5,000 TO 20,000
[12 , 9]CURLY HIGH ALTITUDE CUMULUS

DOWN

[1 , 11]FORM LAYERS LIKE FOG,SPREAD OUT
[2 , 13]COMPOSED OF ICE CRYSTALS,CURLY
[5 , 15]DANGEROUS TO PILOTS,DARK THUNDERSTORMS

15

```
. . C . . . C I R R O S T R A T U S A .
. . . U . . . . . . . . . . . . L .
. . . . M C I R R O C U M U L U S . T .
. . . . U . . . . . . . . . S O .
. . . . . L A L T O S T R A T U S C .
. . . . . O . . . . . R . . . U .
. C . . . N . . . . R . . . M .
. . U . . . I . . . I . . . U .
. . . M . . . M . C . . . . L .
. . . U . . . . B . . . U .
S . . . L . . . U . . . S .
. T . . . U . . . S .
. . R . . . . S .
. . A .
. . . T .
. . . . U .
. . . . . S .
```

THE ANSWERS
ACROSS
[1 , 1]ALTOCUMULUS
[2 , 13]CIRROSTRATUS
[6 , 10]CUMULUS
[9 , 14]ALTOSTRATUS
[12 , 9]CIRROCUMULUS

DOWN
[1 , 11]STRATUS
[2 , 13]CIRRUS
[5 , 15]CUMULONIMBUS

17

NAME: Clouds

SKILL: Science

PROCEDURE: Study the various cloud symbols. Match the symbol with its corresponding meaning. Crayons can be used to write directly on the laminated surface of the folder. Erase your answers after you check them.

VARIATIONS: 1. Study the "Key to Aviation Weather Report" to see how it explains the sky and ceiling and visibility.
2. Put the correct cloud symbol in your daily class calendar.

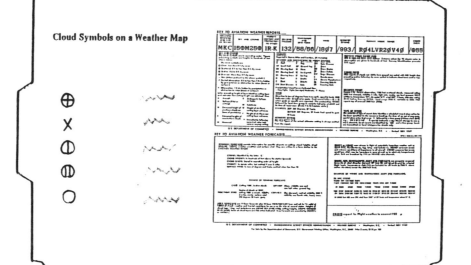

Cloud Symbols on a Weather Map

CLEAR SCATTERED BROKEN OBSCURED OVERCAST

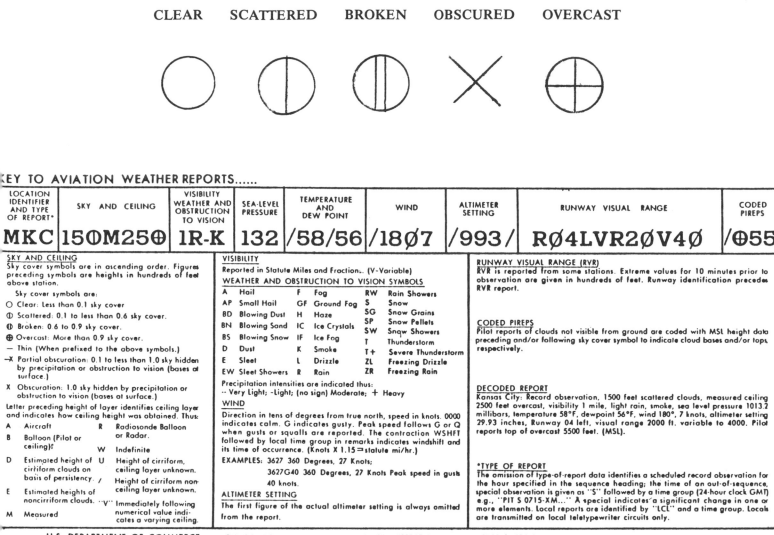

KEY TO AVIATION WEATHER REPORTS......

LOCATION IDENTIFIER AND TYPE OF REPORT*	SKY AND CEILING	VISIBILITY WEATHER AND OBSTRUCTION TO VISION	SEA-LEVEL PRESSURE	TEMPERATURE AND DEW POINT	WIND	ALTIMETER SETTING	RUNWAY VISUAL RANGE	CODED PIREPS
MKC	15⦶M25⊕	1R-K	132	/58/56	/18Ø7	/993/	RØ4LVR2ØV4Ø	/⊕55

SKY AND CEILING

Sky cover symbols are in ascending order. Figures preceding symbols are heights in hundreds of feet above station.

Sky cover symbols are:

O Clear: Less than 0.1 sky cover

⦶ Scattered: 0.1 to less than 0.6 sky cover.

⦶ Broken: 0.6 to 0.9 sky cover.

⊕ Overcast: More than 0.9 sky cover.

— Thin (When prefixed to the above symbols.)

⟋X Partial obscuration: 0.1 to less than 1.0 sky hidden by precipitation or obstruction to vision (bases at surface.)

X Obscuration: 1.0 sky hidden by precipitation or obstruction to vision (bases at surface.)

Letter preceding height of layer identifies ceiling layer and indicates how ceiling height was obtained. Thus:

A Aircraft

B Balloon (Pilot or ceiling)

D Estimated height of cirriform clouds on basis of persistency.

E Estimated heights of noncirriform clouds.

M Measured

R Radiosonde Balloon or Radar.

W Indefinite

U Height of cirriform, ceiling layer unknown.

⟋ Height of cirriform non-ceiling layer unknown.

ᵛ Immediately following numerical value indicates a varying ceiling.

VISIBILITY

Reported in Statute Miles and Fraction.. (V-Variable)

WEATHER AND OBSTRUCTION TO VISION SYMBOLS

A	Hail	F	Fog	RW	Rain Showers
AP	Small Hail	GF	Ground Fog	S	Snow
BD	Blowing Dust	H	Haze	SG	Snow Grains
BN	Blowing Sand	IC	Ice Crystals	SP	Snow Pellets
BS	Blowing Snow	IF	Ice Fog	SW	Snow Showers
D	Dust	K	Smoke	T	Thunderstorm
E	Sleet	L	Drizzle	T+	Severe Thunderstorm
EW	Sleet Showers	R	Rain	ZL	Freezing Drizzle
				ZR	Freezing Rain

Precipitation intensities are indicated thus:
-- Very Light; -Light; (no sign) Moderate; + Heavy

WIND

Direction in tens of degrees from true north, speed in knots. 0000 indicates calm. G indicates gusty. Peak speed follows G or Q when gusts or squalls are reported. The contraction WSHFT followed by local time group in remarks indicates windshift and its time of occurrence. (Knots X 1.15 ≈ statute mi/hr.)

EXAMPLES: 3627 360 Degrees, 27 Knots;

3627G40 360 Degrees, 27 Knots Peak speed in gusts 40 knots.

ALTIMETER SETTING

The first figure of the actual altimeter setting is always omitted from the report.

RUNWAY VISUAL RANGE (RVR)

RVR is reported from some stations. Extreme values for 10 minutes prior to observation are given in hundreds of feet. Runway identification precedes RVR report.

CODED PIREPS

Pilot reports of clouds not visible from ground are coded with MSL height data preceding and/or following sky cover symbol to indicate cloud bases and/or tops respectively.

DECODED REPORT

Kansas City: Record observation, 1500 feet scattered clouds, measured ceiling 2500 feet overcast, visibility 1 mile, light rain, smoke, sea level pressure 1013.2 millibars, temperature 58°F, dewpoint 56°F, wind 180°, 7 knots, altimeter setting 29.93 inches, Runway 04 left, visual range 2000 ft. variable to 4000. Pilot reports top of overcast 5500 feet. (MSL).

*TYPE OF REPORT

The omission of type-of-report data identifies a scheduled record observation for the hour specified in the sequence heading; the time of an out-of-sequence special observation is given as "S" followed by a time group (24-hour clock GMT) e.g., "PIT S 0715-XM..." A special indicates a significant change in one or more elements. Local reports are identified by "LCL" and a time group. Locals are transmitted on local teletypewriter circuits only.

U.S. DEPARTMENT OF COMMERCE • ENVIRONMENTAL SCIENCE SERVICES ADMINISTRATION • WEATHER BUREAU • Washington, D.C.

KEY TO AVIATION WEATHER FORECASTS.......

TERMINAL FORECASTS contain information for specific airports on ceiling, cloud heights, cloud amounts, visibility, weather condition and surface wind. They are written in a form similar to the AVIATION WEATHER REPORT.

CEILING: Identified by the letter "C"

CLOUD HEIGHTS: In hundreds of feet above the station (ground)

CLOUD LAYERS: Stated in ascending order of height

VISIBILITY: In statute miles, but omitted if over 8 miles

SURFACE WIND: In tens of degrees and knots; omitted when less than 10.

EXAMPLE OF TERMINAL FORECASTS

C15⦶ Ceiling 1500', broken clouds

20⦶C70⊕6K 3230G Scattered clouds at 2000', ceiling 7000' overcast, visibility 6 miles, smoke, surface wind 320 degrees 30 knots, gusty.

O11/2GF Clear, visibility one and one-half miles, ground fog.

C5X1/4S+ Sky obscured, vertical visibility 500 ft. visibility one-fourth mile, heavy snow.

AREA FORECASTS are 12-hour forecasts plus 12-hour OUTLOOKS (18 hour outlook in FA valid at 1300Z) of cloud, weather and frontal conditions for an area the size of several states. Heights of cloud tops, icing, and turbulence are ABOVE SEA LEVEL (ASL); ceiling heights, ABOVE GROUND LEVEL (AGL); bases of cloud layers are ASL unless indicated. Area Forecasts are amended by SIGMET's or AIRMET's.

SIGMET or AIRMET warn airmen in flight of potentially hazardous weather such as squall lines, thunderstorms, fog, icing, and turbulence. SIGMET concerns severe and extreme conditions of importance to all aircraft. AIRMET concerns less severe conditions which may be hazardous to some aircraft or to relatively inexperienced pilots. Both are broadcast by FAA on NAVAID voice channels.

WINDS AND TEMPERATURES ALOFT (FD) FORECASTS are computer prepared forecasts of wind direction (nearest 10° true N) and speed (knots) for selected flight levels. Temperatures aloft (°C) are included for all levels (≈2500 ft. above station elevation) except the 3000-foot level.

EXAMPLES OF WINDS AND TEMPERATURES ALOFT (FD) FORECASTS:

FD WBC 121745
BASED ON 121200Z DATA
VALID 130000Z FOR USE 1800-0300Z. TEMPS NEG ABV 24000

FT	3000	6000	9000	12000	18000	24000	30000	34000	39000
BOS	3127	3425-07	3420-11	3421-16	3516-27	3512-38	311649	292451	283451
JFK	3026	3327-08	3324-12	3322-16	3120-27	2923-38	284248	285150	285749

At 6000 feet ASL over JFK wind from 330° at 27 knots and temperature minus 8° C.

PILOTS report in-flight weather to nearest FSS

U.S. DEPARTMENT OF COMMERCE • ENVIRONMENTAL SCIENCE SERVICES ADMINISTRATION • WEATHER BUREAU • Washington, D.C.

CONSTELLATIONS

NAME: Constellations

SKILL: Visual Awareness and Reading Comprehension

PROCEDURE: Read the stories about the Big Dipper and Orion. Study the layout of the stars in the constellations. Use stick-on stars to form the constellations.

VARIATIONS:
1. Look for the constellations outside on a clear night.
2. Visit the library to read about the Greek myths and legends, such as Daedalus, his son Icarus, and the myths of the constellations.
3. Horoscopes are fortunes told by the stars — what is your horoscope sign?

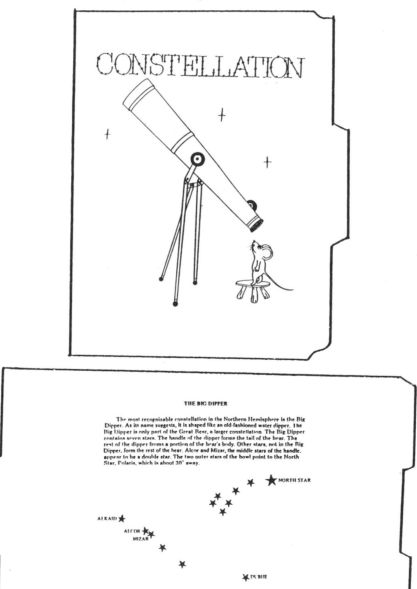

CONSTELLATION

THE BIG DIPPER

The most recognizable constellation in the Northern Hemisphere is the Big Dipper. As its name suggests, it is shaped like an old-fashioned water dipper. The Big Dipper is only part of the Great Bear, a larger constellation. The Big Dipper contains seven stars. The handle of the dipper forms the tail of the bear. The rest of the dipper forms a portion of the bear's body. Other stars, not in the Big Dipper, form the rest of the bear. Alcor and Mizar, the middle stars of the handle, appear to be a double star. The two outer stars of the bowl point to the North Star, Polaris, which is about 30° away.

NORTH STAR

ALKAID

ALCOR
MIZAR

DUBHE

MERAK

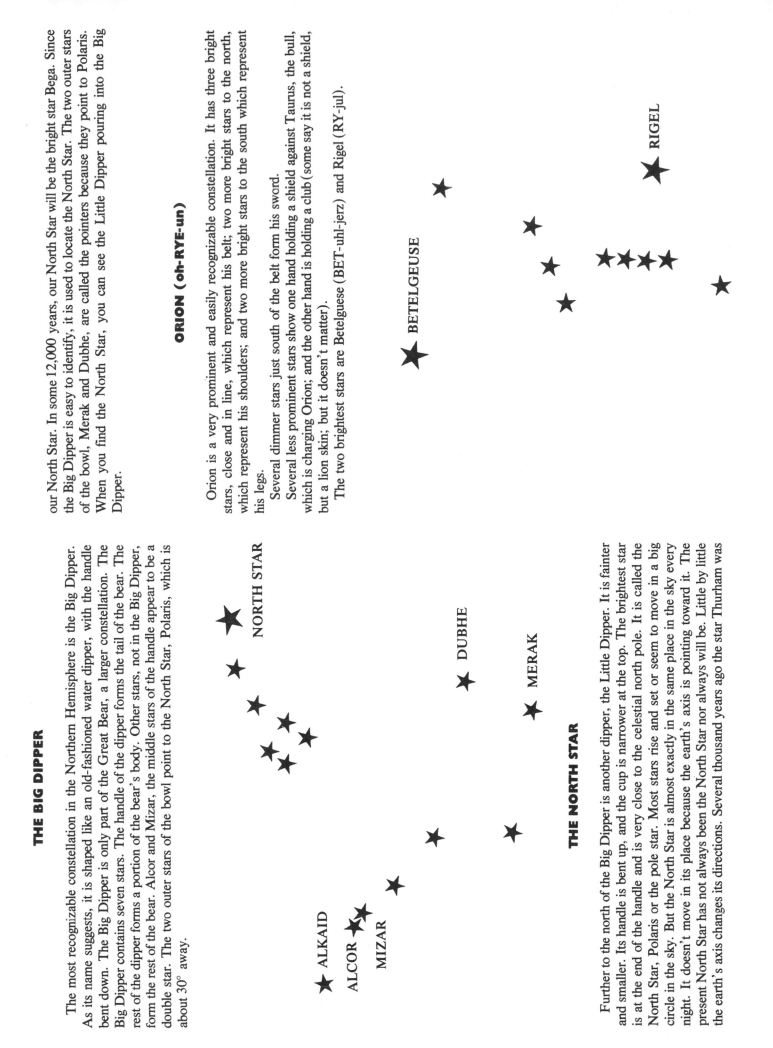

THE BIG DIPPER

The most recognizable constellation in the Northern Hemisphere is the Big Dipper. As its name suggests, it is shaped like an old-fashioned water dipper, with the handle bent down. The Big Dipper is only part of the Great Bear, a larger constellation. The Big Dipper contains seven stars. The handle of the dipper forms the tail of the bear. The rest of the dipper forms a portion of the bear's body. Other stars, not in the Big Dipper, form the rest of the bear. Alcor and Mizar, the middle stars of the handle appear to be a double star. The two outer stars of the bowl point to the North Star, Polaris, which is about 30° away.

our North Star. In some 12,000 years, our North Star will be the bright star Bega. Since the Big Dipper is easy to identify, it is used to locate the North Star. The two outer stars of the bowl, Merak and Dubhe, are called the pointers because they point to Polaris. When you find the North Star, you can see the Little Dipper pouring into the Big Dipper.

ORION (oh-RYE-un)

Orion is a very prominent and easily recognizable constellation. It has three bright stars, close and in line, which represent his belt; two more bright stars to the north, which represent his shoulders; and two more bright stars to the south which represent his legs.

Several dimmer stars just south of the belt form his sword.

Several less prominent stars show one hand holding a shield against Taurus, the bull, which is charging Orion; and the other hand is holding a club (some say it is not a shield, but a lion skin; but it doesn't matter).

The two brightest stars are Betelguese (BET-uhl-jerz) and Rigel (RY-jul).

THE NORTH STAR

Further to the north of the Big Dipper is another dipper, the Little Dipper. It is fainter and smaller. Its handle is bent up, and the cup is narrower at the top. The brightest star is at the end of the handle and is very close to the celestial north pole. It is called the North Star, Polaris or the pole star. Most stars rise and set or seem to move in a big circle in the sky every night. The North Star is almost exactly in the same place in the sky every night. It doesn't move in its place because the earth's axis is pointing toward it. The present North Star has not always been the North Star nor always will be. Little by little the earth's axis changes its directions. Several thousand years ago the star Thurham was

RIGEL

BETELGEUSE

NORTH STAR

DUBHE

MERAK

ALKAID

ALCOR

MIZAR

SOLAR SYSTEM

NAME: Solar System

SKILL: Sequencing Activity

PROCEDURE: Arrange the planets in the correct orbital position.

VARIATIONS: 1. Use a flashlight to represent the sun and styrofoam balls to represent the planets. Rotate the planets slowly on their axes (your hands) and orbit (walk around the sun). Watch how the sun gives the planets light.

2. Learn the poem:

Mercury, Venus, Earth and Mars,
These are the planets that dwell near the stars.
Jupiter, Saturn, Uranus, too, Neptune and Pluto,
I know them — do you?

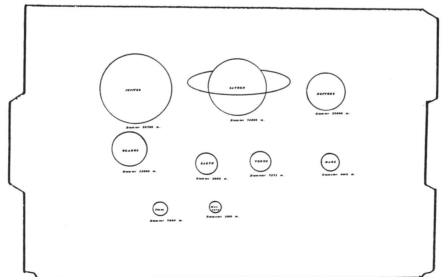

24

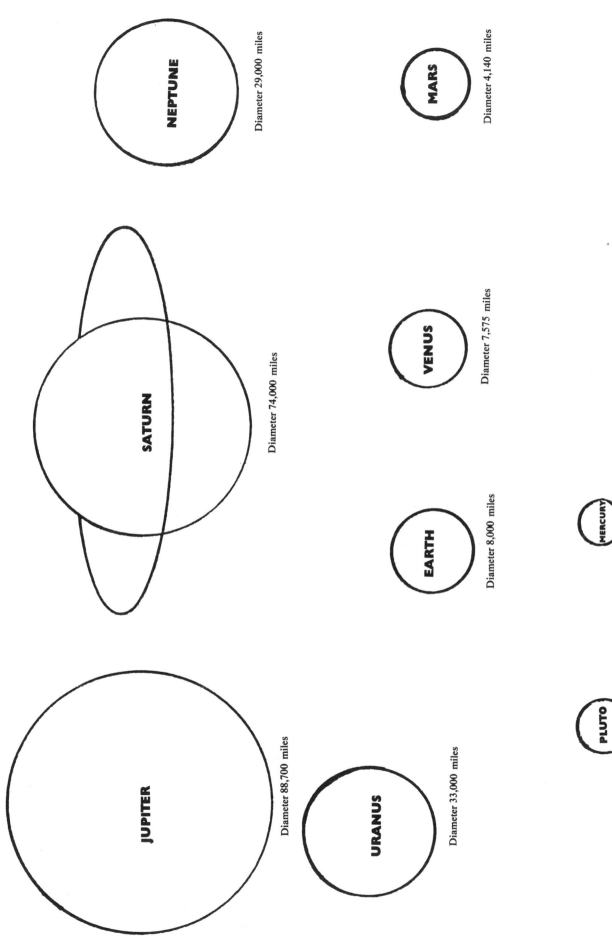

NEPTUNE
Diameter 29,000 miles

MARS
Diameter 4,140 miles

SATURN
Diameter 74,000 miles

VENUS
Diameter 7,575 miles

EARTH
Diameter 8,000 miles

MERCURY
Diameter 3,100 miles

JUPITER
Diameter 88,700 miles

URANUS
Diameter 33,000 miles

PLUTO
Diameter 7,000 miles

planets

I.Q.

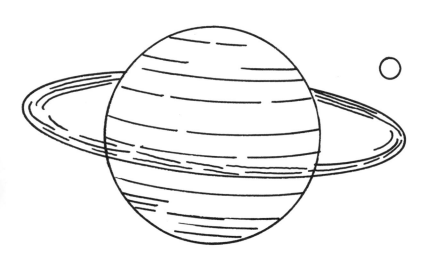

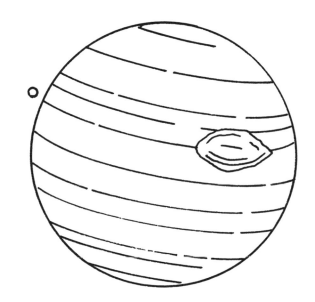

NAME: Planets IQ

SKILL: Reading Comprehension

PROCEDURE: Read the story about the planets and answer the questions. Check your answers before erasing the laminated folder.

VARIATIONS: 1. Tape record the planet stories. Listen to the tape then answer the questions.
2. Write real or imaginary stories about the various planets.
3. Visit the library and find out more about the planets.
4. Make a paper mache sun and planets.

THE PLANETS

Mercury: The planet Mercury is the closest to the sun. It's hard to see but is yellow in color. It doesn't have any moons. The length of each year is 88 days. Life isn't possible because there is no air, water, and it's too hot. It is 36,000,000 miles from the sun. Its diameter is 3100 miles. Mercury was named for the Greek God, Mercury, which means swift messenger carrier.

Venus: The planet closest to the size of the earth is Venus. It seems to change shape and we call it our morning or evening star. It takes 225 days to go around the sun. It is 67,200,000 miles from the sun. The diameter is 7,575 miles. The life, as we know it on earth, isn't possible on Venus. It doesn't have any moons. Venus is named for the God of Beauty.

Earth: The earth is the only planet that has life on it. It has one moon. The length of each day is twenty-four hours. The Earth is the only planet that isn't named for a god. The diameter is 8,000 miles. It is 893 million miles from the sun. It takes 365 1/4 days to revolve around the sun.

Mars: Mars was named for the God of War. It takes two of our years for Mars to go around the sun. The length of each day is four hours and thirty-seven minutes. It has two moons. The diameter is 4,140 miles. It is 141,000,000 miles from the sun.

Two Viking spacecraft landed on Mars on July 20, 1976. It took almost a year and a half a billion miles, but the craft landed safely. The cameras on the Viking craft scanned 1.7 million square miles. They found rolling dunes of orange dust and volcanic rocks. There appeared to be evidence that water may have been on Mars about a billion years ago or the possibility that water is underground. No life was found. However, many interesting facts were found and many questions about the universe were raised.

Jupiter: Jupiter is the largest planet. It was named for the Ruler of the Gods. It has twelve moons. Life isn't possible because it is too cold and there are poisonous gases. The length of each day is about ten hours. It takes twelve years to go around the sun. It is 483,000,000 miles from the sun. The diameter is 88,700 miles.

Man's first trips to Jupiter came in March 1972 and April 1973, with two spacecraft Pioneers 10 and 11. The two craft were the first to go beyond the orbit of Mars, pass through the Asteroid Belt, and to reach Jupiter. The trip covered more than a half-billion miles.

Saturn: It takes Saturn about 29½ years to go around the sun. Saturn has rings and will float on water. It is 887,000,000 miles from the sun. It has nine moons and three rings. The size in diameter is 74,000 miles. Life isn't possible since it's too cold. The length of each day is ten and one-fourth hours. It was named for the Greek God of Harvest.

Uranus: Uranus is green in color and was named for the Greek Grandfather of Gods. It has five moons. Life isn't possible because it's too cold. The length of each day is ten hours and forty-five minutes. The diameter is 29,000 miles. It's 1,786,000,000 miles from the sun. It takes 84 years for Uranus to go around the sun.

Neptune: Neptune's two moons go in opposite directions. It was named for the God of Fresh Water and Sea. It takes 165 years to go around the sun. It is 2,793,000,000 miles from the sun. It is 33,000 miles in diameter. The length of each day is 16 hours.

Pluto: Pluto is the farthest planet from the sun. However, scientists believe they may have discovered another planet even farther! Pluto is named for the God of the Underworld. We don't know if it has any moons. It takes 249 years to go around the sun. The length of each day is six days and nine hours. It is 3,670,000,000 miles from the sun. The diameter is 7,000 miles.

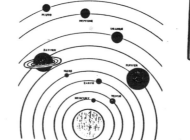

THE PLANETS

We live on a planet called the Earth. This planet travels around the sun. Scientists hope to take trips to and land on the other eight planets which travel around the sun. Four of these planets are larger than the earth and four are smaller. All nine planets receive heat and light from the sun. Planets appear to shine steadily in the sky while stars twinkle.

Mercury: The planet Mercury is the closest to the sun. It's hard to see but is yellow in color. It doesn't have any moons. The length of each year (i.e., the time it takes to orbit around the sun) is 88 days. Life isn't possible because there is no air or water, and it's too hot. It is 36 million miles from the sun. Its diameter is 3,100 miles. Mercury was named for the Greek god, Mercury, which means swift message carrier.

Venus: The planet closest to the size of the Earth is Venus. It seems to change shape and we call it our morning or evening star. It takes 225 days to go around the sun. It is 67,200,000 miles from the sun. The diameter is 7,575 miles. The life, as we know it on Earth, isn't possible on Venus. It doesn't have any moons. Venus is named for the goddess of beauty.

Earth: The earth is the only planet that has life on it. It has one moon. The length of each day is twenty-four hours. Earth is the only planet that isn't named for a god. The diameter is 8,000 miles. It is 93 million miles from the sun. It takes 365¼ days to revolve around the sun.

Mars: Mars was named for the god of war. It takes almost two of our years for Mars to go around the sun. The length of each day is four hours and thirty-seven minutes. It has two moons. The diameter is 4,140 miles. It is 141 million miles from the sun.
Two Viking spacecraft landed on Mars on July 20, 1976. It took almost a year to travel the half a billion miles, but the craft landed safely. The cameras on Viking craft scanned 1.7 million square miles. They found rolling dunes of orange dust and volcanic rocks. There appeared to be evidence that water may have been on Mars about a billion years ago or that water is underground. No life was found. However, many interesting facts were found and many questions about the universe were raised.

Jupiter: Jupiter is the largest planet. It was named for the ruler of the gods. It has twelve moons. Life isn't possible because it is too cold and there are poisonous gases. The length of each day is about ten hours. It takes twelve earth years to go around the sun. It is 483 million miles from the sun. The diameter is 88,700 miles.
Man's first trips to Jupiter came in March 1972 and April 1973, with two spacecraft Pioneers 10 and 11. The two craft were the first to go beyond the orbit of Mars, pass through the Asteroid Belt, and to reach Jupiter. The trip covered more than a half-billion miles.
Many interesting experiments were performed and many exciting photographs of the Polar Region, the Great Red Spot, and the banded cloud tops were taken.

Saturn: It takes Saturn about 29½ earth years to go around the sun. It is 887 million miles from the sun. It has nine moons and three rings. The size in diameter is 74,000 miles. Life isn't possible since it's too cold. The length of each day is 10¼ hours. It was named for the Greek god of harvest.

Uranus: Uranus is green in color and was named for the great grandfather of gods. It has five moons. Life isn't possible because it's too cold. The length of each day is 10¾ hours. The diameter is 29,000 miles. It is 1,786,000,000 miles from the sun. It takes 84 earth years for Uranus to go around the sun.

Neptune: Neptune's two moons go in opposite directions. It was named for the god of fresh water and the sea. It takes 165 earth years to go around the sun. It is 2,793,000,000 miles from the sun. It is 33,000 miles in diameter. The length of each day is 16 hours.

Pluto: Pluto is the farthest planet from the sun. However, scientists believe they may have discovered a planet even farther! Pluto is named for the god of the underworld. We don't know if it has any moons. It takes 249 earth years to go around the sun. The length of each day is six days and nine hours. It is 3,670,000,000 miles from the sun. The diameter is 7,000 miles.

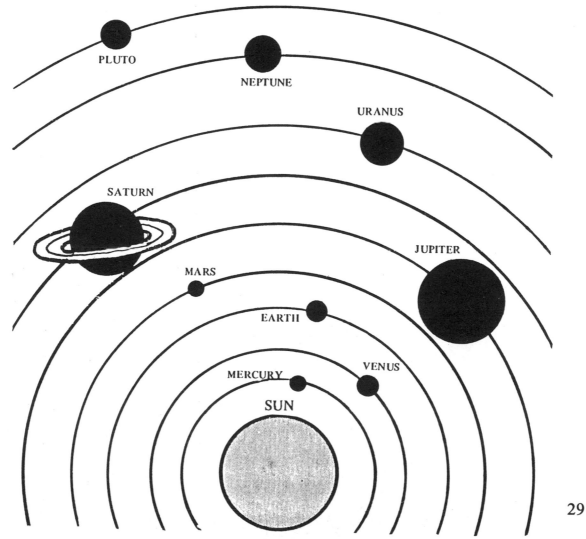

PLANET I. Q.

1. Which planet is the largest? _____

2. Which planet has the most moons? _____

3. Which planet has rings around it? _____

4. How many *days* would it take Jupiter to go around the sun? _____

5. Which is the brightest planet in the sky? _____

6. Which planet has a reddish appearance? _____

7. Which planet is 93 million miles from the sun? _____

8. Which planet could have a temperature of around 700° Fahrenheit on one side? _____

9. Can you list the planets in order according to size from the smallest to the largest?

_____   _____

_____   _____

_____   _____

_____   _____

_____

Answers to "Planet I. Q." — 1) Jupiter, 2) Jupiter, 3) Saturn, 4) 4,380 earth days, 5) Venus, 6) Mars, 7) Earth, 8) Mercury, 9) Mercury, Pluto, Mars, Venus, Earth, Neptune, Uranus, Saturn, Jupiter.

The First

FLIGHT

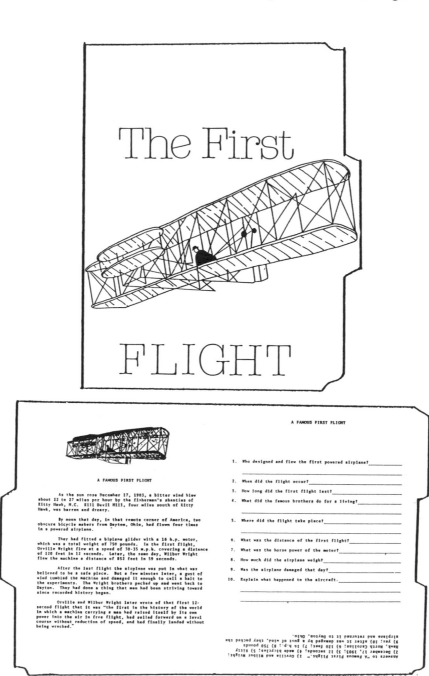

NAME: A Famous First Flight

SKILL: Reading and Comprehension

PROCEDURE: Read the story and answer the questions. Have the teacher check the answers or check your own answers.

VARIATIONS: 1. Tape the story. Listen to the tape, following along with the words. Then answer the questions.
2. Write a newspaper account of the First Flight. Was the public aware of the importance of this flight?

The First

FLIGHT

A FAMOUS FIRST FLIGHT

As the sun rose December 17, 1903, a bitter wind blew about 22 to 27 miles per hour by the fisherman's shanties of Kitty Hawk, N.C. Kill Devil Hill, four miles south of Kitty Hawk, was barren and dreary.

By noon that day, in that remote corner of America, two obscure bicycle makers from Dayton, Ohio, had flown four times in a powered airplane.

They had fitted a biplane glider with a 16 h.p. motor, which was a total weight of 750 pounds. In the first flight, Orville Wright flew at a speed of 30-35 m.p.h. covering a distance of 120 feet in 12 seconds. Later, the same day, Wilbur Wright flew the machine a distance of 852 feet in 59 seconds.

After the last flight the airplane was put in what was believed to be a safe place. But a few minutes later, a gust of wind tumbled the machine and damaged it enough to call a halt to the experiments. The Wright brothers packed up and went back to Dayton. They had done a thing that man had been striving toward since recorded history began.

Orville and Wilbur Wright later wrote of that first 12-second flight that it was "the first in the history of the world in which a machine carrying a man had raised itself by its own power into the air in free flight, had sailed forward on a level course without reduction of speed, and had finally landed without being wrecked."

A FAMOUS FIRST FLIGHT

1. Who designed and flew the first powered airplane?_____
_____
2. When did the flight occur?_____
3. How long did the first flight last?_____
4. What did the famous brothers do for a living?_____
_____
5. Where did the flight take place?_____
_____
6. What was the distance of the first flight?_____
7. What was the horse power of the motor?_____
8. How much did the airplane weigh?_____
9. Was the airplane damaged that day?_____
10. Explain what happened to the aircraft._____
_____
_____

Answers to "A Famous First Flight": 1) Orville and Wilbur Wright; 2) December 17, 1903; 3) 12 seconds; 4) made bicycles; 5) Kitty Hawk, North Carolina; 6) 120 feet; 7) 16 h.p.; 8) 750 pounds 9) yes; 10) after it was damaged by a gust of wind, they packed the airplane and returned it to Dayton, Ohio.

A FAMOUS FIRST FLIGHT

As the sun rose on December 17, 1903, a bitter wind blew about 22 to 27 miles per hour by the fishermen's shanties of Kitty Hawk, North Carolina. Kill Devil Hill, four miles south of Kitty Hawk, was barren and dreary.

By noon that day, in that remote corner of America, two obscure bicycle makers, Orville and Wilbur Wright from Dayton, Ohio, had flown four times in a powered airplane.

They had fitted a biplane with a 16 h.p. motor, which had a total weight of 750 pounds. In the first flight, Orville Wright flew at a speed of 30 – 35 miles per hour, covering a distance of 120 feet in 12 seconds. Later the same day Wilbur Wright flew the machine a distance of 852 feet in 59 seconds.

After the last flight was completed, a gust of wind tumbled the machine and damaged it enough to call a halt to the experiments. The Wright brothers packed up and went back to Dayton. They had done something that man had been striving towards since recorded history began.

Orville and Wilbur Wright later wrote of that first 12-second flight that it was "the first in the history of the world in which a machine carrying a man had raised itself by its own power into the air in free flight, had sailed forward on a level course without reduction of speed, and had finally landed without being wrecked."

A FAMOUS FIRST FLIGHT QUIZ

1. Who designed and flew the first powered airplane? _____
_____

2. When did the flight occur? _____

3. How long did the first flight last? _____

4. What did the famous brothers do for a living? _____
_____

5. Where did the flight take place? _____

6. What was the distance of the first flight? _____

7. What was the horse power of the motor? _____

8. How much did the airplane weight? _____

9. Was the airplane damaged that day? _____

10. Explain what happened to the aircraft. _____
_____
_____

Answers to "A Famous First Flight." — 1) Orville and Wilbur Wright; 2) December 17, 1903; 3) 12 seconds; 4) made bicycles; 5) Kitty Hawk, North Carolina; 6) 120 feet; 7) 16 h.p.; 8) 750 pounds; 9) yes; 10) after it was damaged by a gust of wind, they packed the airplane and returned it to Dayton, Ohio.

AIRCRAFT

HISTORY

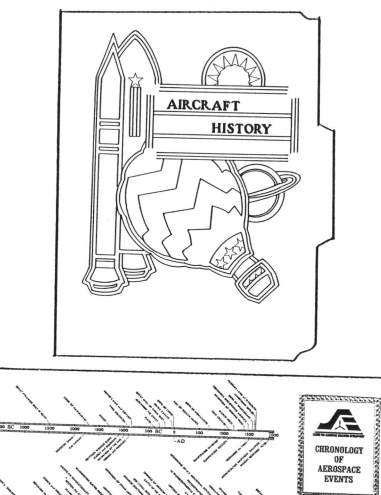

NAME: Aircraft History

SKILL: Library Skills/Reading, Writing, Comprehension

PROCEDURE: Study the Chronology of Aerospace Events. Choose a particular event that interests you. Go to the library to find out more about the event and write a detailed paper about it.

VARIATIONS: 1. Ask your father, grandfather or friend's relatives about events in which they may have participated. Take notes or tape record the interview as they give you an 'oral history' of the event.

2. Continue the time line to the present shuttle launch. Make a class mural of these aircraft.

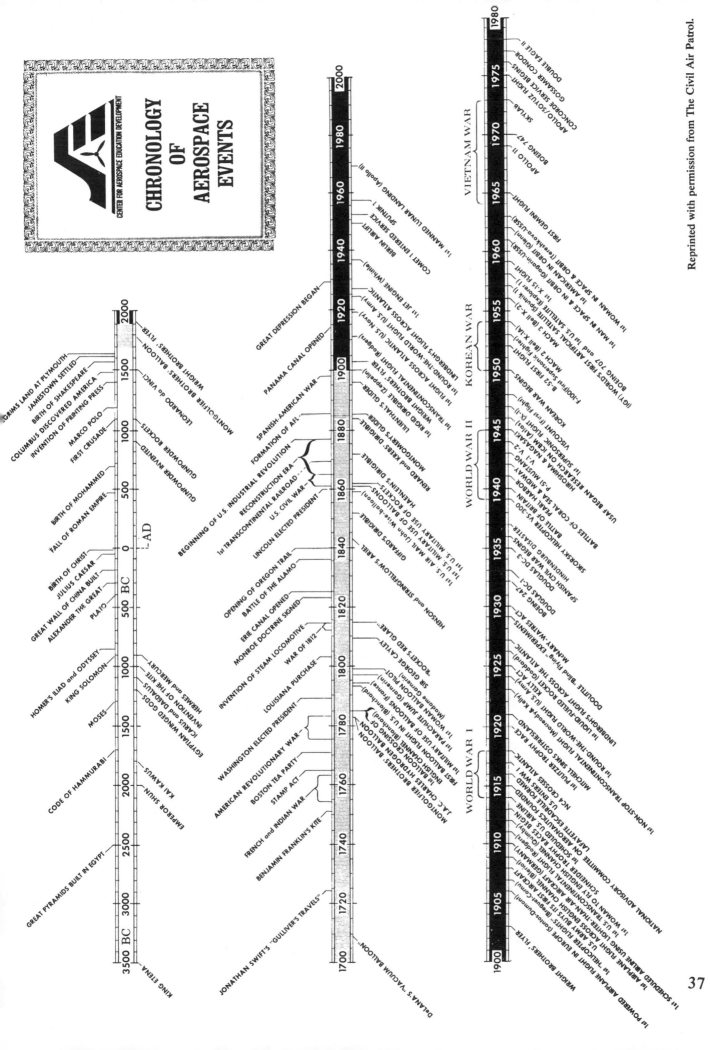

Reprinted with permission from The Civil Air Patrol.

37

EXAMPLE OF ORAL HISTORY

Interview an ex-aviator and see what you can find out about aviation. (Elmer Asbaugh is the father of Eleanor Ashbaugh, a student in Dr. Caballero's aerospace education workshop at the University of Miami).

Orlando Morning Sentinel, November 17, 1936. The "Original Florida Airlines" began with this Ford Trimotor Airplane. "The huge plane brings in two families and furnishings," announced *The Sentinal*.

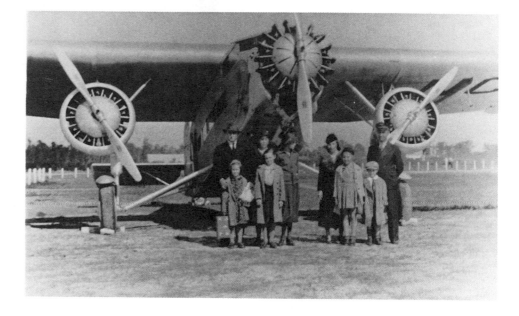

Jeannette News Dispatch, May 24, 1938. **500 AIR LETTERS**

More than 500 airmail letters went out of the Manor post office Friday. Postmaster Homer C. Kifer left the Manor office for Harrison City at 3:10 Friday afternoon with letters which were placed in Elmer Ashbaugh's plane which arrived at Bettis field with thousands of letters from the surrounding district at 3:30.

AIR MAIL PICKUP— John Weightman, Postmaster at Claridge and William Smith, Postmaster at Harrison City, announced yesterday that the air mail will be picked up at the Harrison City field at 3:30 o'clock today. All letters must be at the post office not later than 2:30.

38

Airplane

Flight

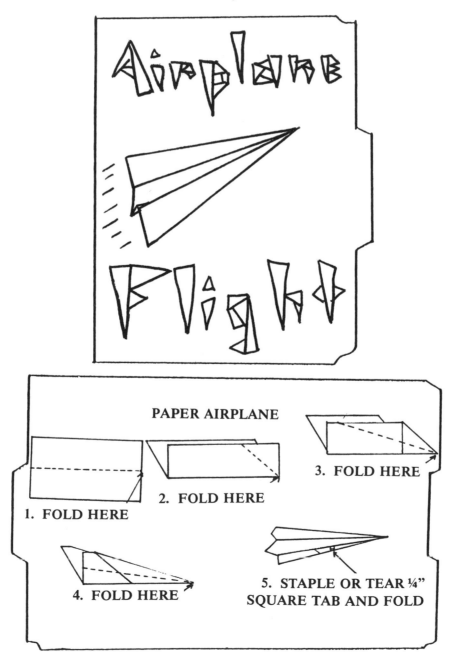

NAME: Airplane Flight

SKILL: Fine Motor Development (demonstration of flight principles).

PROCEDURE: Trace the airplane body, wing and tail onto cardboard. Pull the wing through the slot on the airplane and the tail into the rudder slot. Place a paper clip on the nose. Fly the plane and see what maneuvers you can make.

VARIATIONS: 1. Make the folded paper plane and test fly under the teacher's directions. Experiment with the design.
2. Write for the Delta Dart kits.
3. Visit the library to read various books about airplanes.

PAPER AIRPLANE

1. FOLD HERE

2. FOLD HERE

3. FOLD HERE

4. FOLD HERE

5. STAPLE OR TEAR ¼"
SQUARE TAB AND FOLD

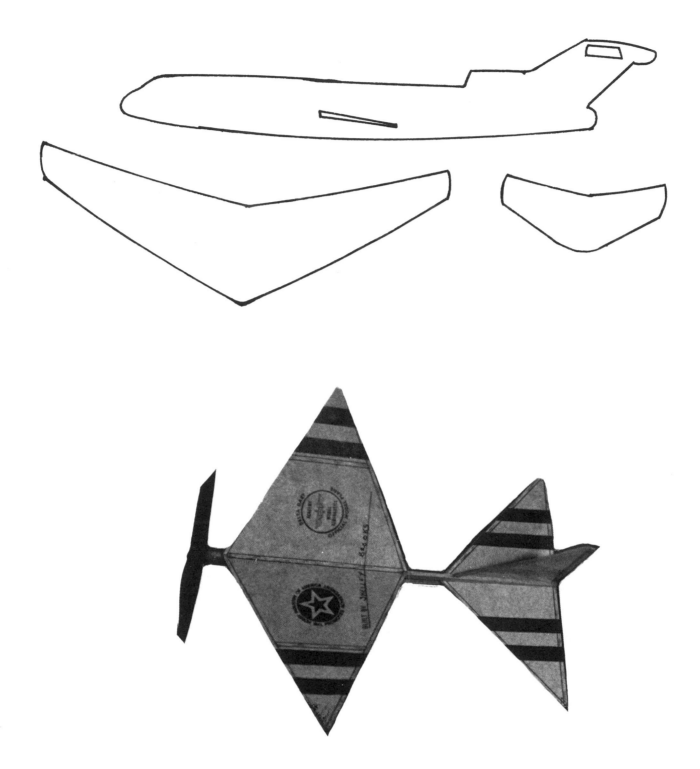

Delta Dart Kits can be purchased in packages of 35 kits including a Teacher's Guide. Write to: Delta Dart Project, Midwest Products Company, 400 S. Indiana Street, Hobart, Indiana 46342.

Aviation Pioneers

NAME: Aviation Pioneers

SKILL: History and Language

PROCEDURE: Play the word search and puzzle game. Check your answers.

VARIATIONS:
1. Using the reader's guide in the library, find out about other aerospace pioneers. Read the following books: Gay, George. *Sole Survivor*; Scott, Sheila. *Barefoot in the Sky* and *The Sky's the Limit;* Yeager, Charles and Leo Janos. *Yeager.*
2. Interview your parents or parents' friends to find out about other aerospace pioneers. Invite them to speak to your class.

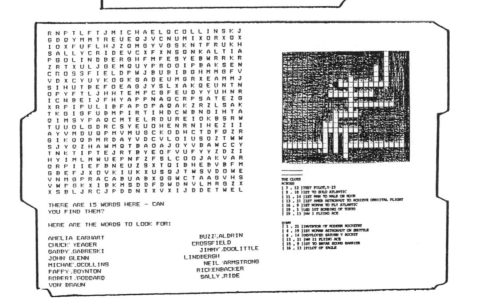

```
R N F T L F T J M I C H A E L Q C O L L I N S K J
G D D Y M M T R E U E Q J V C N U M I X D R X Q X
I O X F U F L H J Z Q M G Y V G S K N T F R U K H
S A L L Y C R I D E V C X F X N S Q N K A L T I A
P G O L I N D B E R G H F M F E S Y E B W R R K R
Z R T X U L J G E M Q U Y P R O Q I P B A K S E N
C R O S S F I E L D F W J B U B I B G H M M G F V
V D X C Y U Y K O G K G A D E U M G R X E A M M J
S I H U T B E F O E A G J Y S L X A K Q E U N T N
O P Y F T L J H H T E M P C G F E U D Y Y U H N R
I C N B E I J F H Y A P F N A Q C R P S A T E Z G
X R P I P U L I B F A P O P A Q A K Z R Z L S A K
T K G I G F U B M P I R T I H D C W B N G I H T A
Q I M S Y P A Q C M T E L R D U R E I O K B S R W
T U U O L G D R C S Y E U O H E N R N I H E Z I I
A Y V M D U Q P M V M U G C K O D H C T D F Q Z R
G I K Q Q R M R D A Y V D C V L O I U S Q Z T W W
S J Y Q Z H A W M Q T B A O A J O Y V B A W C C Y
T N K T I P T E J R T B Y E Q F V U F Y Y Z D Z I
H Y I M L M W U E P N F Z F S L C O O J A K V A R
Q R P I I E F B N E U Z S X T Q I B H E B V B F M
G B E F J X O V K I U K X U S Q J T W S V D O W E
V N M G P R A C A B U A B X Q G W C T A A G V H S
V W P B K X I B K M S D D F D W D N V L M R G Z X
X S B L J R C J P D D N X X V X I J D D E T W E L
```

THERE ARE 15 WORDS HERE — CAN
YOU FIND THEM?

HERE ARE THE WORDS TO LOOK FOR:

AMELIA EARHART
CHUCK YEAGER
GABBY GABRESKI
JOHN GLENN
MICHAEL COLLINS
PAPPY BOYNTON
ROBERT GODDARD
VON BRAUN

BUZZ ALDRIN
CROSSFIELD
JIMMY DOOLITTLE
LINDBERGH
NEIL ARMSTRONG
RICKENBACKER
SALLY RIDE

THE CLUES
ACROSS
7 , 13 |TEST PILOT,X-15
9 , 10 |1ST TO SOLO ATLANTIC
11 , 14 |1ST MAN TO WALK ON MOON
13 , 11 |1ST AMER ASTRONAUT TO ACHIEVE ORBITAL FLIGHT
16 , 9 |1ST WOMAN TO FLY ATLANTIC
19 , 1 |LED 1ST BOMBING OF TOKYO
21 , 13 |WW I FLYING ACE

DOWN
1 , 21 |INVENTOR OF MODERN ROCKETRY
4 , 19 |WOMAN ASTRONAUT ON SHUTTLE
6 , 16 |DEVELOPED SATURN V ROCKET
13 , 11 |WW II FLYING ACE
15 , 9 |1ST TO BREAK SOUND BARRIER
16 , 13 |PILOT OF EAGLE

43

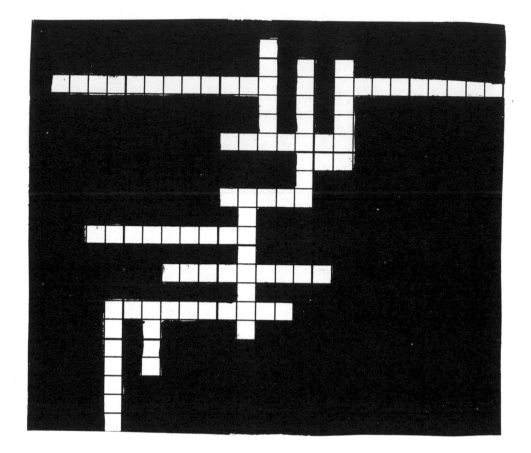

THE CLUES
ACROSS
[7 , 12]TEST PILOT,X-15
[9 , 10]1ST TO SOLO ATLANTIC
[11 , 14]1ST MAN TO WALK ON MOON
[13 , 11]1ST AMER ASTRONAUT TO ACHIEVE ORBIITAL FLIGHT
[16 , 9]1ST WOMAN TO FLY ATLANTIC
[19 , 1]LED 1ST BOMBING OF TOKYO
[19 , 13]WW I FLYING ACE

DOWN
[1 , 21]INVENTOR OF MODERN ROCKETRY
[4 , 19]1ST WOMAN ASTRONAUT ON SHUTTLE
[6 , 14]DEVELOPED SATURN V ROCKET
[13 , 11]WW II FLYING ACE
[15 , 9]1ST TO BREAK SOUND BARRIER
[16 , 13]PILOT OF EAGLE

```
R N P T L F T J M I C H A E L Q C O L L I N S K J
G D Q Y M M T R E U E Q J V C N U M I X O R X Q X
I O X F U F L H J Z Q M G Y V G S K N T F R U K H
S A L L Y C R I D E V C X F X N S Q N K A L T I A
P G O L I N D B E R G H F M F E S Y E B W R R K R
Z R T X U L J G E M Q U Y P R O O I P B A K S E N
C R O S S F I E L D P W J B U B I Q G H M M Q F V
V D X C Y U Y K O G K G A D E U M G R X E A M M J
S I H U T B E F O E A G J Y S L X A K Q E U N T N
O P Y F T L J H H T E M P C G F E U D Y Y U H N R
I C N B E I J F H Y A P P N A Q C R P S A T E Z G
X R P I P U L I B F A P O P A Q A K Z R Z L S A K
T K G I G F U B M P I R T I H D C W B N G I H T A
Q I M S Y P A Q C M T E L R D U R E I O K B S R W
T U U O L G D R C S Y E U O H E N R N I H E Z I I
A Y V M D U Q P M V M U G C K O D H C T D F Q Z R
G I K Q Q B M R D A Y V D C V L O I U S O Z T W W
S J Y Q Z H A W M Q T B A O A J O Y V B A W C C Y
T N K T I P T E J R T B Y E O F V U F Y Y Z D Z I
H Y I M L M W U E P N F Z F S L C O O J A K V A R
Q R P I I E F B N E U Z S X T Q I B H E B V B F M
G B E F J X O V K I U K X U S Q J T W S V D O W E
V N M G P R A C A B U A B X Q G W C T A A G V H S
V W P G K X I B K M S D D F D W D N V L M R G Z X
X S B L J R C J P D D N X X V X I J D D E T W E L
```

There are 15 words here — can you find them?

HERE ARE THE WORDS TO LOOK FOR:

Amelia Earhart

Buzz Aldrin

Chuck Yeager

Crossfield

Gabby Gabreski

Jimmy Doolittle

John Glenn

Lindbergh

Michael Collins

Neil Armstrong

Pappy Boynton

Rickenbacker

Robert Goddard

Sally Ride

Von Braun

THE ANSWERS
ACROSS
[7,12] CROSSFIELD
[9,10] LINDBERGH
[11,14] ARMSTRONG
[13,11] GLENN
[16,9] EARHART
[19,1] DOOLITTLE
[19,13] RICKENBACKER

DOWN
[1,21] GODDARD
[4,19] RIDE
[6,14] VON BRAUN
[13,11] GABRESKI
[15,9] YEAGER
[16,13] ALDRIN

```
. . . . . . . . M I C H A E L Q C O L L I N S . .
. . . . . . . . . . . . . . . . . . I . O . . . .
. . . . . . . . . . . . . . . . . K . T . . . . .
S A L L Y C R I D E . . . . . . S . N . . . T . .
. . . L I N D B E R G H . . . E . Y . . . R . . R
. . . . . . . . . . . . . . R . O . . . A . . E .
C R O S S F I E L D . . . B . B . . . H . G . . .
. . . . . . . . . . . A . E . . . R . . A . N . .
. . . . . . . . . . E . G . Y . . A . . E . N . N
. . . . . . . J . . . Y . P . N . Q . R P . U . N .
. . . . . . . . I B . A . O . A . A K . R . L . . .
. . . . . . . . B M P . R . I . D C . B N G . . . .
. . . . . . . A . . M T . L . D U R E I Q . . . . .
. . . . . . G . . . S Y E . O H E N R N . . . . . .
. . . . . . . . M . M U G C K O D H . . . . . . . .
. . . . . . . R . A . V D C V L O . . . . . . . . .
. . . . . . A . . . T . A Q A J . . . . . . . . . .
. . . . . P . . . R . B . E O . . . . . . . . . . .
. . . . L . . . E . N . Z . . L . . . . . . . . . .
. . . . I . . . B . E . Z . . . . I . . . . . . . .
. . E . . O . K . U . . . . . . . T . . . . . . . .
. N . . . R . C . B . . . . . . . . . . T . . . . .
. . . . . . . I . . . . . . . . . . . . L . . . . .
. . . . . . . R . . . . . . . . . . . . . E . . . .
```

46

Theory of FLIGHT

NAME: Theory of Flight

SKILL: Reading Comprehension

PROCEDURE: 1. Read about Bernoulli's law and perform the experiment. Draw your own airplane, noting where the forces occur.

VARIATIONS: 1. Visit a Civil Air Patrol squadron to find out about opportunities relating to aviation
2. Contact the Air Force recruiter to find out about the Air Force.

THEORY OF FLIGHT

What makes an airplane fly,
And keeps it up in the clear, blue sky?
Four forces of flight — lift, drag, weight, and thrust,
Get to know them; a good pilot must!

The important principle to understand as to why an airfoil can produce lift is to consider *Bernoulli's law*. He proved that where the speed of a moving gas is high, the pressure is low. Where the speed is low, the pressure is high.

A simple experiment will help you see how it works to produce *lift*.

Cut a piece of paper two inches wide and seven inches long. Hold it against the chin under your bottom lip with the narrow part. Then blow hard over the top of the paper. The paper rises!

What actually happens is the "air in a hurry" on top of the paper has less pressure. The pressure under the paper is greater and lifts the paper up.

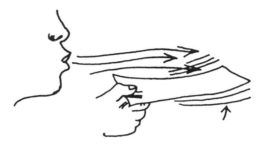

If you take the same paper and just pull it through the air, it will rise again.

The wing of an airplane rises when it is pulled through the air by an engine just as the paper is pushed up by greater pressure below. The air moving over the curved wing on top must travel faster to reach the back of the wing. Some of the air goes under the wing also, but they reach the trailing edge at the same time. Therefore, the air pressure on top of the wing is less than the pressure on the bottom of the wing, so the plane lifts up.

LIFT

Low pressure area

High pressure area

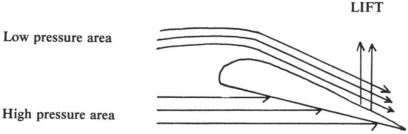

Level Flight and Constant Speed

Center of Gravity, CG is the balance point of the aircraft. The CG must be maintained within design limits to have proper control of the aircraft.

Thrust is equal to the *drag* and acts below the CG to cause a slight lifting action to the nose which mostly overcomes the pitch down tendency caused by the lift acting behind the CF.

Lift is equal to weight and acts behind CG so that aircraft will nose down when power is reduced. Tail load could be up or down depending on the location of the CG and the effect of the combined forces.

Weight always acts toward earth. The direction of the other forces depends on the position of the aircraft.

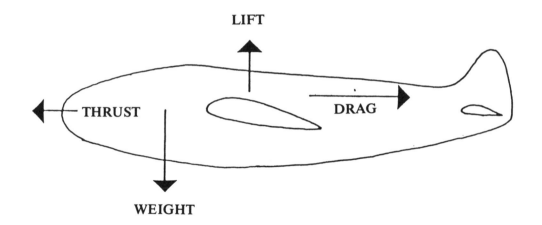

If you are interested in being a pilot, start thinking about the Civil Air Patrol. The CAP is a civilian organization and an auxiliary of the United States Air Force. The minimum age requirement is thirteen or you must be in the sixth grade. You can study many topics such as military leadership, aerospace education, and first aid and survival. Encampments are also provided.

An Air Force pilot can be a very interesting and exciting career. It is very expensive to receive proper training. It can cost over two million dollars for five years of training. (The Air Force Academy, pilot training, upgrading to other aircraft and for flight missions).

Below are the approximate costs of operating planes per hour:

C 130	$1,546
F 4	$2,275
C 141	$3,119
B 52	$5,414
C 5	$7,763

AIRPLANE

PARTS

NAME: Airplane Parts

SKILL: Vocabulary Development

PROCEDURE: Study the illustrations. Cover up your answers to see how many parts you can name.

VARIATIONS:
1. Obtain a basic book on flying to learn how the controls and instruments work. Explain the functions of the parts.
2. Visit a local airport and ask a pilot to show you the parts of the airplane.
3. Compare the different planes — the outside parts and the instrument panels inside.

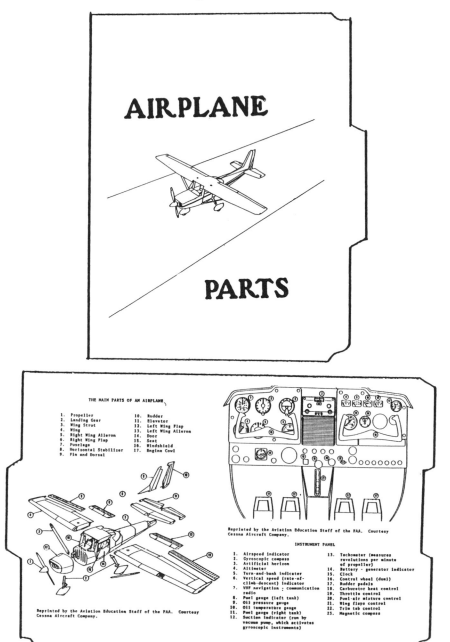

AIRPLANE

PARTS

THE MAIN PARTS OF AN AIRPLANE

1. Propeller
2. Landing Gear
3. Wing Strut
4. Wing
5. Right Wing Aileron
6. Right Wing Flap
7. Fuselage
8. Horizontal Stabilizer
9. Fin and Dorsal
10. Rudder
11. Elevator
12. Left Wing Flap
13. Left Wing Aileron
14. Door
15. Seat
16. Windshield
17. Engine Cowl

Reprinted by the Aviation Education Staff of the FAA. Courtesy Cessna Aircraft Company.

INSTRUMENT PANEL

1. Airspeed indicator
2. Gyroscopic compass
3. Artificial horizon
4. Altimeter
5. Turn-and-bank indicator
6. Vertical speed (rate-of-climb-descent) indicator
7. VHF navigation - communication radio
8. Fuel gauge (left tank)
9. Oil pressure gauge
10. Oil temperature gauge
11. Fuel gauge (right tank)
12. Suction indicator (run by vacuum pump, which activates gyroscopic instruments)
13. Tachometer (measures revolutions per minute of propeller)
14. Battery - generator indicator
15. Clock
16. Control wheel (dual)
17. Rudder pedals
18. Carburetor heat control
19. Throttle control
20. Fuel-air mixture control
21. Wing flaps control
22. Trim tab control
23. Magnetic compass

Reprinted by the Aviation Education Staff of the FAA. Courtesy Cessna Aircraft Company.

THE MAIN PARTS OF AN AIRPLANE

1. Propeller
2. Landing Gear
3. Wing Strut
4. Wing
5. Right Wing Aileron
6. Right Wing Flap
7. Fuselage
8. Horizontal Stabilizer
9. Fin and Dorsal

10. Rudder
11. Elevator
12. Left Wing Flap
13. Left Wing Aileron
14. Door
15. Seat
16. Windshield
17. Engine Cowl

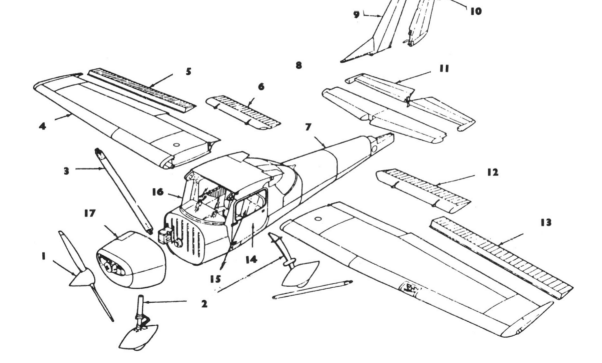

INSTRUMENT PANEL

1. Airspeed indicator
2. Gyroscopic compass
3. Artificial horizon
4. Altimeter
5. Turn-and-bank indicator
6. Vertical speed (rate-of-climb-descent) indicator
7. VHF navigation — communication radio
8. Fuel gauge (left tank)
9. Oil pressure gauge
10. Oil temperature gauge
11. Fuel gauge (right tank)

12. Suction indicator (run by vacuum pump, which activites gyroscopic instruments)
13. Tachometer (measures revolutions per minute of propeller)
14. Battery — generator indicator
15. Clock
16. Control wheel (dual)
17. Rudder pedals
18. Carburetor heat control
19. Throttle control
20. Fuel-air mixture control
21. Wing flaps control
22. Trim tab control
23. Magnetic compass

Reprinted by the Aviation Education Staff of the FAA. Courtesy Cessna Aircraft Company.

Planning a flight

NAME: Planning a Flight

SKILL: Reading and Mathematics skills

PROCEDURE: 1. Obtain an aeronautical chart from a local airport or your State Department of Transportation and compare how it is different from a local road map.
2. Find and explain the symbols on your aeronautical chart. Study the airman's phonetic alphabet.

VARIATIONS: 1. Develop the skills learned in these experiences and use them on an introductory flight at a small flight school (if your age permits!).
2. Invite a pilot to your classroom to talk about learning to fly and how to fill out a flight plan.

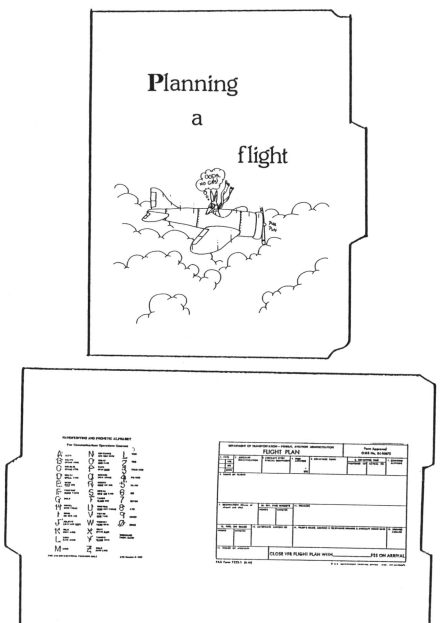

Operator Department   Student Handout
KTTC, Mississippi   1 February 1967

HANDPRINTING AND PHONETIC ALPHABET

For Communications Operations Courses

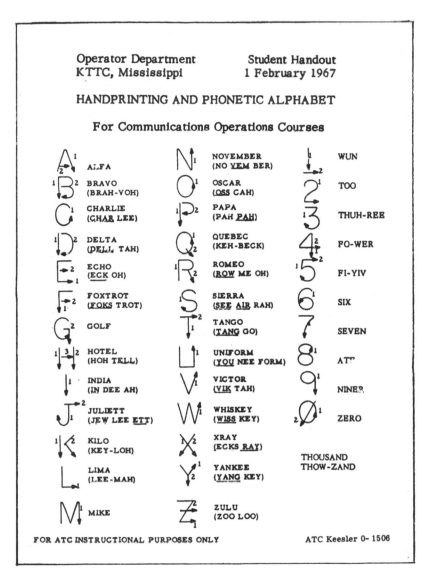

ALFA	NOVEMBER (NO VEM BER)	WUN
BRAVO (BRAH-VOH)	OSCAR (OSS CAH)	TOO
CHARLIE (CHAR LEE)	PAPA (PAH PAH)	THUH-REE
DELTA (DELL TAH)	QUEBEC (KEH-BECK)	FO-WER
ECHO (ECK OH)	ROMEO (ROW ME OH)	FI-YIV
FOXTROT (FOKS TROT)	SIERRA (SEE AIR RAH)	SIX
GOLF	TANGO (TANG GO)	SEVEN
HOTEL (HOH TELL)	UNIFORM (YOU NEE FORM)	ATE
INDIA (IN DEE AH)	VICTOR (VIK TAH)	NINER
JULIETT (JEW LEE ETT)	WHISKEY (WISS KEY)	ZERO
KILO (KEY-LOH)	XRAY (ECKS RAY)	
LIMA (LEE-MAH)	YANKEE (YANG KEY)	THOUSAND THOW-ZAND
MIKE	ZULU (ZOO LOO)	

DEPARTMENT OF TRANSPORTATION— FEDERAL AVIATION ADMINISTRATION

FLIGHT PLAN

Form Approved
OMB No. 04-R0072

1. TYPE	2. AIRCRAFT IDENTIFICATION	3. AIRCRAFT TYPE/ SPECIAL EQUIPMENT	4. TRUE AIRSPEED	5. DEPARTURE POINT	6. DEPARTURE TIME		7. CRUISING ALTITUDE
VFR					PROPOSED (Z)	ACTUAL (Z)	
IFR							
DVFR			KTS				

8. ROUTE OF FLIGHT

9. DESTINATION (Name of airport and city)	10. EST. TIME ENROUTE		11. REMARKS
	HOURS	MINUTES	

12. FUEL ON BOARD		13. ALTERNATE AIRPORT (S)	14. PILOT'S NAME, ADDRESS & TELEPHONE NUMBER & AIRCRAFT HOME BASE	15. NUMBER ABOARD
HOURS	MINUTES			

16. COLOR OF AIRCRAFT	CLOSE VFR FLIGHT PLAN WITH_____FSS ON ARRIVAL

FAA Form 7233-1 (5-72)

U.S. GOVERNMENT PRINTING OFFICE

PILOT'S PREFLIGHT CHECK LIST

WORLD AERONAUTICAL CHARTS

SCALE 1:1,000,000

Nautical Miles / Statute Miles

CLOSE FLIGHT PLAN UPON ARRIVAL

WEATHER ADVISORIES		ALTERNATE WEATHER		DATE
EN ROUTE WEATHER		FORECASTS	NOTAMS	
DESTINATION WEATHER		WINDS ALOFT	AIRSPACE RESTRICTIONS	
		MAPS		

FLIGHT LOG

	VOR				DISTANCE	TIME			GROUND SPEED
	IDENT.	TO	LEG	PT-TO-PT CUMULATIVE		TAKEOFF			
	FREQ.	FROM	REMAINING			ETA	ATA		
DEPARTURE POINT	RADIAL								
CHECK POINT									
DESTINATION									
TOTAL									

POSITION REPORT: FVFR report hourly, IFR as required by ATC

ACFT. IDENT.	POSITION	TIME	ALT.	IFR/VFR	EST. NEXT FIX	NAME OF SUCCEEDING FIX	PIREPS

REPORT CONDITIONS ALOFT— CLOUD TOPS, BASES, LAYERS, VISIBILITY, TURBULENCE, HAZE, ICE, THUNDERSTORMS

SCALE 1:500,000

SECTIONAL AERONAUTICAL CHARTS

Nautical Miles / Statute Miles

59

NAME: A Trip to the Airport

SKILL: Reading Comprehension (Spanish and English)

PROCEDURE: Write for the publication:
A Trip to the Airport
Federal Aviation Administration
800 Independence Ave., SW
Washington, D.C. 20591.

VARIATIONS:
1. Read the story in Spanish to make the students aware of a second language. If they speak Spanish, encourage them to translate the story.
2. Visit a local airport. Find out how many airports (small and large) are within a 30 mile radius of your home. How many jobs are available because of these airports?

61

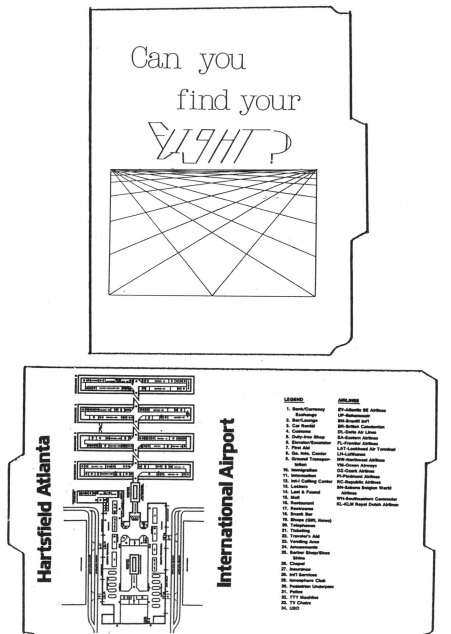

NAME: Can You Find Your Flight?

SKILL: Reading and Comprehension

PROCEDURE: Cut out the airline ticket jacket (bend on dotted line). Insert the ticket and boarding pass. Read the flight information to see which airline you'll be traveling on. Read the legend on the airport map at Hartsfield Atlanta International Airport to locate the concourse and gate number.

VARIATIONS:
1. Make maps of your local airport.
2. Keep your old airline tickets or obtain one from a friend, if possible.
3. Find the flight at your airport. Which airlines serve your airport?

Can you find your FLIGHT?

Hartsfield Atlanta International Airport

LEGEND
1. Bank/Currency Exchange
2. Bar/Lounge
3. Car Rental
4. Customs
5. Duty-free Shop
6. Elevator/Escalator
7. First Aid
8. Ga. Info. Center
9. Ground Transportation
10. Immigration
11. Information
12. Int'l Calling Center
13. Lockers
14. Lost & Found
15. Mail
16. Restaurant
17. Restrooms
18. Snack Bar
19. Shops (Gift, News)
20. Telephones
21. Ticketing
22. Traveler's Aid
23. Vending Area
24. Amusements
25. Barber Shop/Shoe Shine
26. Chapel
27. Insurance
28. Int'l Services
29. Ionosphere Club
30. Pedestrian Underpass
31. Police
32. TTY Machine
33. TV Chairs
34. USO

AIRLINES
EV-Atlantis SE Airlines
UP-Bahamasair
BN-Braniff Int'l
BR-British Caledonian
DL-Delta Air Lines
EA-Eastern Airlines
FL-Frontier Airlines
LAT-Lockheed Air Terminal
LH-Lufthansa
NW-Northwest Airlines
VM-Ocean Airways
OZ-Ozark Airlines
PI-Piedmont Airlines
RC-Republic Airlines
SN-Sabena Belgian World Airlines
WH-Southeastern Commuter
KL-KLM Royal Dutch Airlines

Hartsfield Atlanta

International Airport

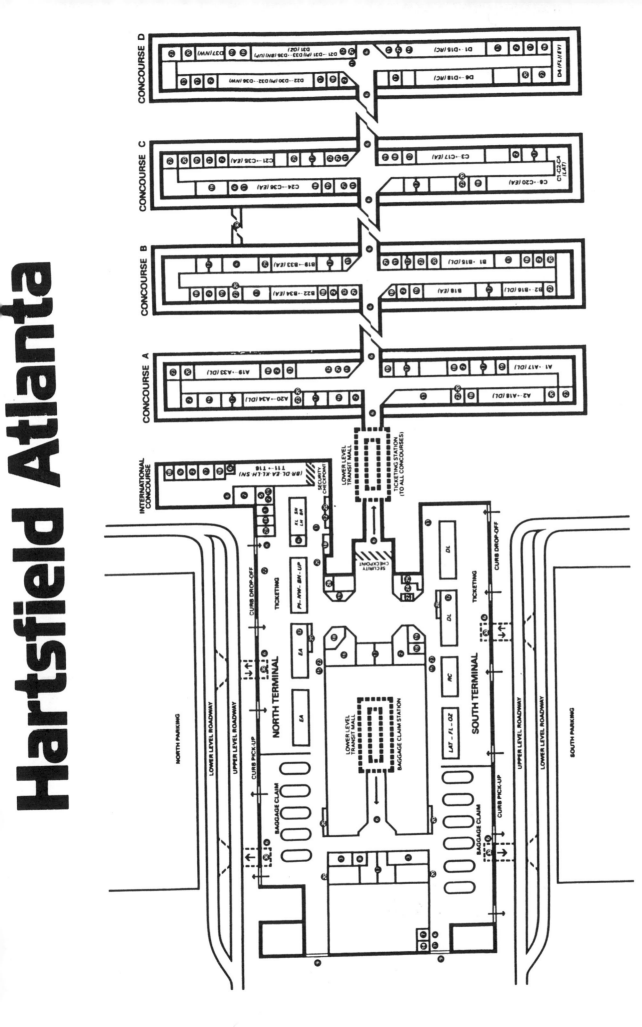

SOLD SUBJECT TO CONDITIONS OF CONTRACT ON PASSENGER'S COUPON

Issued By **NORTHWEST ORIENT**

PASSENGER TICKET AND BAGGAGE CHECK
PASSENGER'S COUPON

012 5844 110 955

ENDORSEMENTS (CARBON)

DATE OF ISSUE

DATE AND PLACE OF ISSUE

NAME OF PASSENGER — NOT TRANSFERABLE

BARKER/JACK

ORIGIN

DESTINATION

COUPONS NOT VALID BEFORE			
1	2	3	4

COUPONS NOT VALID AFTER			
1	2	3	4

ORIGINALLY ISSUED AGAINST BY AGENTS NUMERIC CODE AT ON DATE YR.

TICKET DESIGNATOR & TOUR CODE — THIS TICKET ISSUED IN EXCHANGE FOR

NOT GOOD FOR PASSAGE	FARE BASIS	CARRIER	FLIGHT/CLASS	DATE	TIME	STATUS
ATLANTA	VDG	DL	245V	10JAN	858A	OK
MEMPHIS TENN						
VOID	VOID					
VOID	VOID					
VOID						

BAGGAGE CHECKED UNCK'D. PCS. WT.

FARE	TAX
78.70	
TAX 6.30	TOTAL 85.00

EQUIV. FARE PD.

TRANSMISSION CONTROL NUMBER

FROM / TO: ATL, MEM DL — CARRIER — FARE CALCULATION

Agent 98930440

CONJUNCTION TICKET(S)

FORM OF PAYMENT

ENDORSEMENTS (Carbon)

FARE

CPN | AIRLINE | TICKET NUMBER | CK

012 5844110955 4

REV 6-83 PRINTED IN U.S.A. BY RAND McNALLY

DELTA® AIR LINES

NOTE: FOR SECURITY REASONS, ALL UNCHECKED ARTICLES CARRIED IN THE CABIN ARE SUBJECT TO SEARCH.

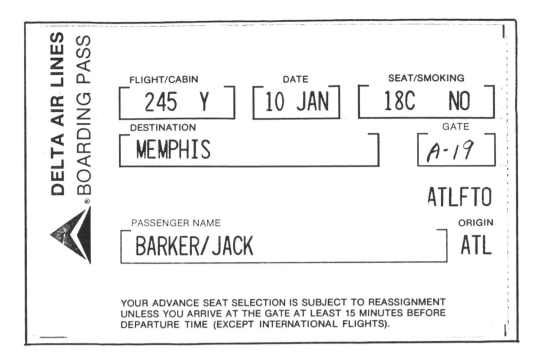

DELTA AIR LINES
® BOARDING PASS

FLIGHT/CABIN	DATE	SEAT/SMOKING
245 Y	10 JAN	18C NO

DESTINATION
MEMPHIS

GATE
A-19

ATLFTO

PASSENGER NAME
BARKER/JACK

ORIGIN
ATL

YOUR ADVANCE SEAT SELECTION IS SUBJECT TO REASSIGNMENT
UNLESS YOU ARRIVE AT THE GATE AT LEAST 15 MINUTES BEFORE
DEPARTURE TIME (EXCEPT INTERNATIONAL FLIGHTS).

LEGEND

1. Bank/Currency Exchange
2. Bar/Lounge
3. Car Rental
4. Customs
5. Duty-free Shop
6. Elevator/Escalator
7. First Aid
8. Ga. Info. Center
9. Ground Transportation
10. Immigration
11. Information
12. Int-l Calling Center
13. Lockers
14. Lost & Found
15. Mail
16. Restaurant
17. Restrooms
18. Snack Bar
19. Shops (Gift, News)
20. Telephones
21. Ticketing
22. Traveler's Aid
23. Vending Area
24. Amusements
25. Barber Shop/Shoe Shine
26. Chapel
27. Insurance
28. Int'l Services
29. Ionosphere Club
30. Pedestrian Underpass
31. Police
32. TTY Machine
33. TV Chairs
34. USO

AIRLINES

EV-Atlantic SE Airlines
UP-Bahamasair
BN-Braniff Int'l
BR-British Caledonian
DL-Delta Air Lines
EA-Eastern Airlines
FL-Frontier Airlines
LAT-Lockheed Air Terminal
LH-Lufthansa
NW-Northwest Airlines
VM-Ocean Airways
OZ-Ozark Airlines
PI-Piedmont Airlines
RC-Republic Airlines
SN-Sabena Belgian World Airlines
WH-Southeastern Commuter
KL-KLM Royal Dutch Airlines

AIRPLANE

NAME: Airplane/Hangar

SKILL: Fine Motor Skill Coordination

PROCEDURE: Complete the worksheets. Complete the airplane and help the airplane get to the hangar.

VARIATIONS:
1. Draw your own mazes for the airplane to follow.
2. Draw other aircraft, i.e. the Shuttle, helicopter, balloons, and create your own connect-the-dot picture.

67

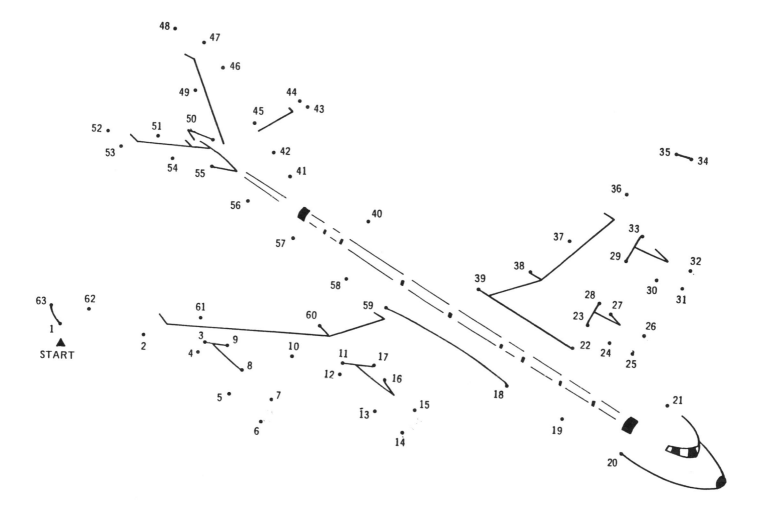

48
47
46
49 44
45 43
50
52 51
53 42
54 55 41
56
57 40
58
63 62 61
1 60 59
START 2 3 9
4 10 11 17
8 12 16
5 7 13 15 18
6 14

35 34
36
33
37 32
29
38 30 31
39 28 27
23 26
22 24
25 21
19
20

69

NAME: Aviation Spoken Here

SKILL: Mathematics

PROCEDURE: Read the airline system schedule and answer the questions.

VARIATIONS: 1. Obtain time tables from other airlines and compare their schedules. (Individual airline schedules will be considerably easier to read for lower grades).

2. Plan a trip to a friend's or relative's home. Find out what flights go there and which is the most economical.

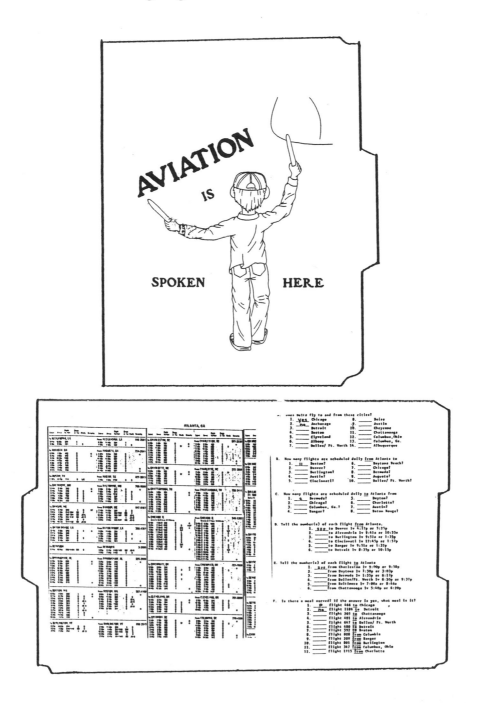

To CHICAGO, ILL. CST CHI

C-CGX (MEIGS FIELD)
M-MDW (MIDWAY)
O-ORD (O'HARE)
P-PWK (PAL-WAUKEE)

From GULFPORT/BILOXI, MISS.-CONT.

Freq.	Leave	Arrive	Flight	Class	Eq	Ml	S
X7	10:55a	11:25a MSY	RC 270	CYBM	D9S		0
	12:45P MSY	2:49P O	NW 183	FYBQV	727	L	0
7	12:40P	2:59P ATL	IW 372	YB	DC9	L	0
	4:05P ATL	4:50P O	DL 806	FYBMV	767	S	0
	1:53P	2:45P MEM	RC 35	YBM	DC9		0
	3:15P MEM	4:40P O	RC 473	YBMK	DC9		0
X6	6:10P	7:50P ATL	RC 119	YBM	DC9		1
	8:40P ATL	9:20P O	RC 279	CYBMK	D9S		0

GUNNISON, COLORADO MST GUC
FARES: F 491.00 Y 229.00-314.00 Q 199.00

Freq.	Leave	Arrive	Flight	Class	Eq	Ml	S
	10:05a	10:50a DEN	VJ 401	YBQ	SWM		0
	11:45a DEN	3:00P O	FL 803	FYBQV	727	L	0
	2:25P	3:04P DEN	VJ 413	YBQ	SWM		0
	3:44P DEN	6:59P O	UA 956	FYBQM	D10	D	0
	4:53P	6:34P DEN	FL 723	YBQMV	CVR		1
	7:10P DEN	10:25P O	FL 807	FYBQV	727	D	0
	7:00P	7:45P DEN	VJ 235	YBQ	SWM		0
	8:25P DEN	11:26P O	UA 278	FYBQM	72S		0

HALIFAX, N.S. AST YHZ
FARES: F 449.00+ Y 281.00+

Freq.	Leave	Arrive	Flight	Class	Eq	Ml	S
6	7:00a	8:17a YYZ	AC 603	FCYB	72S	B	0
	10:20a YYZ	10:46a O	AA 117	FYBQM	72S	S	0
X6	7:20a	8:37a YYZ	AC 603	FCYB	72S	B	0
	10:20a YYZ	10:46a O	AA 117	FYBQM	72S	S	0
X6	12:15P	1:32P YYZ	AC 127	FCYB	72S	L	0
	5:20P YYZ	5:55P O	AC 725	FCYB	72S	D	0
X6	5:05P	6:22P YYZ	AC 627	FCYB	72S	D	0
	8:08P YYZ	8:35P O	AA 293	FYBQM	72S	S	0
6	5:25P	6:45P YYZ	PV 177	CYB	737	D	0
	8:08P YYZ	8:35P O	AA 293	FYBQM	72S	S	0

HANCOCK, MICH. EST CMX
FARES: C 165.00

Freq.	Leave	Arrive	Flight	Class	Eq	Ml	S
X7	7:25a	7:54a MSP	RC 921	YBM	CVR		1
	8:35a MSP	9:40a O	RC 597	CYBMK	D9S	B	0
X6	4:42P	4:56P MSP	RC 925	YBM	CVR		0
	6:05P MSP	7:15P O	RC 658	CYBMK	D9S	S	0

HARLINGEN, TEXAS CST HRL
FARES: F 481.00 Y 296.00
EX: 299.00-419.00

Freq.	Leave	Arrive	Flight	Class	Eq	Ml	S
	9:04a	12:44P O	AA 452	FYBQM	727	B	1

CONNECTIONS
FARES: F 370.00-481.00 Y 296.00-323.00

Freq.	Leave	Arrive	Flight	Class	Eq	Ml	S
	7:35a	8:59a DFW	AA 197	FYBQM	72S	S	0
	9:48a DFW	11:51a O	AA 280	FYBQM	72S	B	0
	12:40P	2:04P DFW	AA 294	FYBQM	727		0
	2:45P DFW	4:55P O	AA 360	FYBQM	767	S	0
X6	12:40P	2:04P DFW	AA 294	FYBQM	727		0
E-15APR	4:00P DFW	6:05P M	ML 352	YM	D9S	S	0
6	12:40P	2:04P DFW	AA 294	FYBQM	727		0
E-15APR	4:00P DFW	6:05P M	ML 352	MK	D9S	S	0
	3:09P	4:33P DFW	AA 328	FYBQM	727		0
	5:13P DFW	7:22P O	AA 564	FYBQM	767	D	0
X6	3:09P	4:33P DFW	AA 328	FYBQM	727		0
	6:30P DFW	8:35P M	ML 342	MK	D9S	S	0

HARRISBURG, PA. EST MDT
FARES: F 362.00 Y 238.00 M 190.00
EX: 296.00-324.00

Freq.	Leave	Arrive	Flight	Class	Eq	Ml	S
	8:26a	9:21a O	UA 637	FYBQM	72S	B	0
X67	10:50a	12:59P O	AL 99	YBM	D9S	S	1
X6	1:50P	3:34P O	AL 169	YBM	D9S		1
	7:40P	8:34P O	UA 459	FYBQM	73S	D	0

CONNECTIONS
FARES: F 507.01 Y 201.00-343.00 M 190.00

Freq.	Leave	Arrive	Flight	Class	Eq	Ml	S
X6	9:58a	11:02a STL	TW 217	FYBM	727	S	0
	11:40a STL	12:42P O	TW 464	FYBM	72S	S	0
	9:58a	11:02a STL	TW 217	FYBM	727	S	0
	1:00P STL	2:02P O	TW 134	FYBM	727	S/	0
X7	10:50a	11:34a PIT	AL 99	YBM	D9S		0
	1:40P PIT	1:59P O	AL 387	YBM	D9S	S	0
	11:00a	11:45a PHL	AL 1202	YBM	SH6		0
E-15APR	12:30P PHL	1:28P M	ML 371	MK	D9S	S	0
	11:00a	11:45a PHL	AL 1202	YBM	SH6		0
D-14APR	1:00P PHL	1:58P M	ML 71	QM	D9S	S	0
	2:50P	4:00P JFK	AL 1208	YBM	SH6		0
	7:10P JFK	8:40P O	CL 219	Y	D8S	S	0

HARRISON, ARK. CST HRO
FARES: F 251.00-277.01 Y 198.01-287.00

Freq.	Leave	Arrive	Flight	Class	Eq	Ml	S
X67	6:50a	8:10a MEM	GM 471	Y	SWM		1
	8:38a MEM	10:00a O	DL 110	FYBMV	72S	S	0
X67	6:50a	8:10a MEM	GM 471	Y	SWM		1
	8:30a MEM	10:00a O	RC 471	CYBMK	D95	B	0
6	6:50a	8:50a MEM	GM 466	Y	SWM		2
	11:10a MEM	12:30P O	RC 475	CYBMK	D95	S	0
X6	12:30P	1:20P MEM	GM 432	Y	SWM		0
	3:15P MEM	4:40P O	RC 473	YBMK	DC9		0
X6	4:20P	6:40P STL	GM 423	Y	SWM		2
	7:15P STL	8:10P O	DL 1538	FYBMV	D9S	S	0
X6	4:20P	6:40P STL	GM 423	Y	SWM		2
	8:00P STL	9:02P O	TW 398	FYBM	72S	S/	0

HARTFORD CT/SPRINGFIELD,MA EST BDL
FARES: F 382.00 Y 255.00
EX: 349.00

Freq.	Leave	Arrive	Flight	Class	Eq	Ml	S
	7:10a	8:22a O	AA 515	FYBQM	727	B	0
	7:55a	9:03a O	UA 103	FYBQM	D10	B	0
	9:35a	10:47a O	AA 291	FYBQM	72S	S	0
	11:00a	12:08P O	UA 463	FYBQM	D10	L	0
	1:10P	2:20P O	AA 269	FYBQM	727	L	0
	1:45P	2:53P O	UA 873	YBQM	72S	S	0
	4:25P	5:36P O	UA 433	FYBQM	72S	S	0
	7:54P	9:02P O	UA 599	FYBQM	767		0

CONNECTIONS
FARES: P 449.99 F 268.00-449.99 Y 202.00-331.00
M 253.00

Freq.	Leave	Arrive	Flight	Class	Eq	Ml	S
	7:25a	7:59a BOS	DL 1161	FYBMV	L10		0
	9:35a BOS	11:00a O	NW 287	FYBQV	72S	S	0
X67	7:30a	8:55a CLE	IW 742	YB	DC9	B	0
D-14APR	9:40a CLE	9:40a M	ML 333	YBM	DC9	S	0
X67	7:30a	8:55a CLE	IW 742	YB	DC9	B	0
E-15APR	9:40a CLE	9:40a M	ML 133	YBM	DC9	S	0
	10:20a	11:00a EWR	PI 423	YBHQ	73S		0
	2:00P EWR	3:17P O	TW 187	FYBM	72S	S	0
X67	12:00n	1:30P DTW	IW 753	YB	DC9	L	0
D-14APR	2:30P DTW	2:20P M	ML 17	QM	DC9		0
X67	12:00n	1:30P DTW	IW 753	YB	DC9	L	0
E-15APR	2:30P DTW	2:20P M	ML 317	MK	D9S		0
X6	12:10P	12:59P LGA	QO 1759	YBQL	BE9		0
	2:50P LGA	4:12P O	TW 337	FYBM	727	S	0
1234	2:30P	4:09P DTW	AL 183	YBM	B11		0
D-14APR	5:05P DTW	4:55P M	ML 7	QM	D9S		0
X6	2:30P	4:09P DTW	AL 183	YBM	B11		0
E-15APR	5:05P DTW	4:55P M	ML 107	YM	DC9		0
57	2:30P	4:09P DTW	AL 183	YBM	B11		0
D-14APR	5:05P DTW	4:55P M	ML 7	MK	D9S		0
X67	2:45P	3:30P BOS	QO 1768	YBQL	BE9		0
	5:55P BOS	7:22P O	NW 35	FCYBQ	72S	D	0
6	3:10P	3:55P BOS	QO 1768	YBQL	BE9		0
	5:55P BOS	7:22P O	NW 35	FCYBQ	72S	D	0
	3:30P	4:15P JFK	PA 946	YBM	F28		0
	6:30P JFK	8:00P O	PA 77	PJYBM	L15	D	0
			PA 946 JOINT OPERATION PA-UR				
	3:30P	4:15P JFK	PA 946	YBM	F28		0
	6:30P JFK	8:08P O	TW 745	FCYBM	767	D	0
			PA 946 JOINT OPERATION PA-UR				
	3:30P	4:15P JFK	PA 946	YBM	F28		0
	7:10P JFK	8:40P O	CL 219	Y	D8S	S	0
			PA 946 JOINT OPERATION PA-UR				
1234	4:20P	6:25P DTW	IW 756	YB	DC9	S	1
D-14APR	7:10P DTW	7:00P M	ML 11	QM	DC9		0

CONT. NEXT COLUMN

To CHICAGO, ILL. CST CHI

From HARTFORD CT/SPRINGFIELD,MA-CONT.

Freq.	Leave	Arrive	Flight	Class	Eq	Ml	S
X67	4:20P	6:25P DTW	IW 756	YB	DC9	S	1
E-15APR	7:10P DTW	7:00P M	ML 311	MK	D9S		0
5	4:20P	6:25P DTW	IW 756	YB	DC9	S	1
D-14APR	7:10P DTW	7:00P M	ML 11	MK	DC9		0
X6	4:56P	6:20P CLE	UA 797	FYBQM	72S	S	0
E-15APR	6:50P CLE	6:50P M	ML 137	YM	DC9	S	0
				EX 16APR			
X6	6:20P	7:23P BUF	AL 451	YBM	B11	S	0
	8:12P BUF	8:40P O	AA 119	FYBQM	727		0
6	7:15P	8:00P JFK	PA 944	YBM	F28		0
D-17APR	8:45P JFK	10:05P O	PA 226	FDYBM	D10	S	0
			PA 944 JOINT OPERATION PA-UR				
6	7:15P	8:00P JFK	PA 944	YBM	F28		0
E-18APR	8:45P JFK	10:05P O	PA 226	FDYBM	72S	S	0
			PA 944 JOINT OPERATION PA-UR				

HAVANA, CUBA EST HAV

Freq.	Leave	Arrive	Flight	Class	Eq	Ml	S
156	9:30a	11:15a MEX	CU 464	Y	TU5		0
	1:10P MEX	6:30P O	MX 802	Y	D10		1

HAVRE, MONT. MST HVR
FARES: F 365.00 Y 290.00

Freq.	Leave	Arrive	Flight	Class	Eq	Ml	S
X7	10:15a	11:45a BIL	GQ 521	YBM	CNA		1
	12:45P BIL	3:59P O	NW 72	FYBQV	72S	L	0

HAYS, KAN. CST HYS
FARES: F 377.00-402.00 Y 199.99

Freq.	Leave	Arrive	Flight	Class	Eq	Ml	S
X7	5:35a	6:46a ICT	ZV 820	YBK	SWM		2
	9:05a ICT	10:44a O	UA 758	FYBQM	73S	S	0
X7	5:35a	8:37a MCI	ZV 820	YBK	SWM		4
	11:47a MCI	12:59P O	TW 52	FCYBM	767	S/	0
X67	11:04a	12:44P MCI	ZV 214	YBK	SWM		1
D-14APR	2:25P MCI	3:35P M	ML 150	QM	DC9		0
X67	11:04a	12:44P MCI	ZV 214	YBK	SWM		1
	2:45P MCI	3:56P O	UA 640	FYBQM	72S		0
7	12:21P	1:40P MCI	ZV 754	YBK	SWM		0
D-14APR	2:25P MCI	3:35P M	ML 150	QM	DC9		0
7	12:21P	1:40P MCI	ZV 754	YBK	SWM		0
	2:45P MCI	3:56P O	UA 640	FYBQM	72S		0
7	12:21P	1:40P MCI	ZV 754	YBK	SWM		0
E-15APR	3:30P MCI	4:40P M	ML 150	MK	DC9	S	0
7	12:21P	1:40P MCI	ZV 754	YBK	SWM		0
	5:32P MCI	6:51P O	TW 454	FYBM	72S	S	0
X6	3:32P	5:05P MCI	ZV 185	YBK	SWM		1
	6:00P MCI	7:17P O	AA 480	FYBQM	727	S	0

HELENA, MONT. MST HLN
FARES: F 406.00-441.00 Y 294.00

Freq.	Leave	Arrive	Flight	Class	Eq	Ml	S
	6:40a	9:09a SLC	WA 460	YBQM	73S		2
	10:15a SLC	1:59P O	WA 370	FYBQM	D10	L	0
	8:39a	10:49a DEN	FL 502	YBQMV	73S		1
	11:45a DEN	3:00P O	FL 803	FYBQV	727	L	0
	11:30a	12:09P BIL	NW 68	FYBQV	72S		0
	12:45P BIL	3:59P O	NW 72	FYBQV	72S	L	0
	2:45P	4:35P SLC	WA 270	YBQM	73S		1
	6:40P SLC	10:24P O	WA 116	FYBQM	72S	D	0

HIBBING, MINN. CST HIB
FARES: C 160.00 Y 134.00-145.00
EX: 259.00-279.00

Freq.	Leave	Arrive	Flight	Class	Eq	Ml	S
X7	10:48a	1:35P O	RC 504	CYBM	D9S	S	2

CONNECTIONS
FARES: F 233.00-258.00 C 160.00-203.00
Y 134.00-183.01

Freq.	Leave	Arrive	Flight	Class	Eq	Ml	S
X7	6:33a	7:43a MSP	RC 72	CYBM	D95		1
	8:35a MSP	9:40a O	RC 597	CYBMK	D9S	B	0
6	6:33a	7:43a MSP	RC 72	CYBM	D95		1
	8:45a MSP	9:53a M	ML 328	MK	D9S	S	0
X67	6:33a	7:43a MSP	RC 72	CYBM	D95		1
D-14APR	8:45a MSP	9:53a M	ML 328	YM	M80	S	0
X67	6:33a	7:43a MSP	RC 72	CYBM	D95		1
E-15APR	8:45a MSP	9:53a M	ML 328	YM	D9S	S	0
X67	8:30a	9:30a MSP	XJ 152	YB	BE9		0
	10:50a MSP	11:59a M	NW 804	FYBQV	727	S	0
X7	10:48a	11:55a MSP	RC 504	CYBM	D9S		1
	1:25P MSP	2:33P M	ML 382	MK	D9S	S	0
X6	3:10P	4:10P MSP	XJ 154	YB	BE9		0
	5:00P MSP	6:10P O	RC 647	CYBMK	D95	S	0
X6	3:10P	4:10P MSP	XJ 154	YB	BE9		0
E-15APR	5:15P MSP	6:23P M	ML 378	YM	D9S	S	0
X6	3:10P	4:10P MSP	XJ 154	YB	BE9		0
D-14APR	5:20P MSP	6:28P M	ML 316	YM	M80	S	0
X6	3:10P	4:10P MSP	XJ 154	YB	BE9		0
	6:55P MSP	8:04P M	NW 818	FYBQV	727	S	0
X6	4:15P	5:00P MSP	RC 166	YBM	CVR		0
	6:05P MSP	7:15P O	RC 658	CYBMK	D9S	S	0
	4:15P	5:00P MSP	RC 166	YBM	CVR		0
	6:04P MSP	7:17P O	UA 492	FYBQM	73S	S	0

HICKORY, N.C. EST HKY
FARES: F 282.00 S 191.99 Y 191.99-244.00

Freq.	Leave	Arrive	Flight	Class	Eq	Ml	S
X67	7:05a	7:30a CLT	ED 100	YBHK	BE9		0
	8:55a CLT	9:45a O	PI 273	YBHQ	73S	B	0
X7	7:15a	7:40a CLT	SG 401	Y	PAG		0
	8:05a CLT	8:48a O	EA 212	FYBML	72S	B	0
X67	10:59a	11:24a CLT	ED 51	YBHK	CNA		0
	12:14P CLT	12:59P O	PI 308	YBHQ	73S	L	0
	11:00a	11:25a CLT	SG 403	Y	PAG		0
	11:50a CLT	12:31P O	EA 218	FYBML	D9S	L	0
X67	2:20P	2:45P CLT	SG 493	Y	PAG		0
	3:15P CLT	4:05P O	PI 342	YBHQ	73S	S	0
X6	3:25P	3:49P CLT	SG 405	Y	PAG		0
	4:43P CLT	5:28P O	EA 706	FYBML	72S	D	0
X6	3:25P	3:50P CLT	ED 104	YBHK	BE9		0
	5:15P CLT	6:05P O	PI 288	YBHQ	73S	D	0
X67	6:10P	6:35P CLT	ED 55	YBHK	CNA		0
	7:19P CLT	8:09P O	PI 284	YBHQ	73S	D	0
X6	6:25P	6:48P CLT	SG 407	Y	PAG		0
	7:19P CLT	8:09P O	PI 284	YBHQ	73S	D	0

HILO, HAWAII; HAWAII HST ITO
FARES: F 690.00-791.11+ Y 537.00-673.56+

Freq.	Leave	Arrive	Flight	Class	Eq	Ml	S
	12:55P	1:35P HNL	HA 241	YQM	D95		0
	5:40P HNL	5:35a O	UA [illegible]	FYQM	747	DS	0
	12:55P	1:35P HNL	HA 241	YQM	D95		0
D-27APR	4:35P HNL	5:53a O	AA 160	*	D10	DS	1
			AA 160 FYBQM-SFO-FnYnBQM				
	12:55P	1:35P HNL	HA 241	YQM	D95		0
	6:00P HNL	5:55a O	UA 10	FYQM	747	DS	0
	1:20P	2:00P HNL	AQ 243	FYQM	73S		0
D-27APR	4:35P HNL	5:53a O	AA 160	*	D10	DS	1
			AA 160 FYBQM-SFO-FnYnBQM				
	3:35P	4:15P HNL	AQ 45	FYQM	73S		0
	5:40P HNL	5:35a O	UA 2	FYQM	747	DS	0
	5:20P	12:09a LAX	UA 301	FYQM	D10	D	0
	12:40a LAX	6:25a O	UA 118	FYBQM	D10	S	0
	6:15P	6:55P HNL	AQ 47	FYQM	73S		0
	11:00P HNL	1:59P O	WA 370	FYBQM	D10	*	2
				EX 28APR			
			WA 370 * MEALS SBL/SL				
	8:20P	9:20P HNL	HA 701	TKL	DH7		0
	11:00P HNL	1:59P O	WA 370	FYBQM	D10	*	2
				EX 28APR			
			WA 370 * MEALS SBL/SL				

HILTON HEAD ISLAND, S.C. EST HHH
FARES: F 328.01 S 238.00 Y 238.00-290.00

Freq.	Leave	Arrive	Flight	Class	Eq	Ml	S
	10:25a	11:25a CLT	SG 233	Y	PA1		0
	11:50a CLT	12:31P O	EA 218	FYBML	D9S	L	0
	2:40P	3:45P CLT	SG 235	Y	PA1		0
	4:43P CLT	5:26P O	EA 706	FYBML	72S	D	0
	5:36P	6:48P CLT	SG 237	Y	PAG		0
	7:19P CLT	8:09P O	PI 284	YBHQ	73S	D	0

HOBBS, N.M. MST HOB
FARES: F 541.00-563.00 Y 397.00

Freq.	Leave	Arrive	Flight	Class	Eq	Ml	S
X7	5:51a	7:30a ABQ	ZV 420	YBK	SWM		1
	9:05a ABQ	12:41P O	AA 372	FYBQM	727	B	0
X7	10:01a	11:40a ABQ	ZV 533	YBK	SWM		1
	3:30P ABQ	9:02P O	TW 398	FYBM	72S	S	1
7	10:01a	11:40a ABQ	ZV 719	YBK	SWM		1
	3:30P ABQ	9:02P O	TW 398	FYBM	72S	S	1

To CHICAGO, ILL. CST CHI

From HONOLULU, OAHU; HAWAII HST HNL
FARES: F 680.00-744.86+ Y 559.00-634.72+
EX: 601.87-630.86+

Freq.	Leave	Arrive	Flight	Class	Eq	Ml	S
	2:40P	4:55a O	UA 116	FYQM	D10	DS	1
	4:35P	5:53a O	AA 160	*	D10	DS	1
			AA 160 DISCONTINUED AFTER 27APR				
			AA 160 FYBQM-SFO-FnYnBQM				
SPEC	4:35P	6:31a O	AA 160	*	D10	DS	1
				OP 28APR			
			AA 160 FYBQM-SFO-FnYnBQM				
	5:40P	5:35a O	UA 2	FYQM	747	DS	0
	6:00P	5:55a O	UA 10	FYQM	747	DS	0
	11:00P	1:59P O	WA 370	FYBQM	D10	*	2
				EX 28APR			
			WA 370 * MEALS SBL/SL				

CONNECTIONS
FARES: F 568.36-972.81+ Y 526.37-746.63+

Freq.	Leave	Arrive	Flight	Class	Eq	Ml	S
37	2:15a	9:20a LAX	CO 2	PCYBM	D10	B	0
	11:07a LAX	4:55P O	AA 194	FYBQM	D10	L	0
14	2:40a	9:45a LAX	CO 2	PCYBM	D10	B	0
	11:07a LAX	4:55P O	AA 194	FYBQM	D10	L	0
	8:45a	3:21P SFO	PA 124	PJYBM	747	B	0
	4:15P SFO	10:05P O	UA 130	FYBQM	D10	D	0
	9:15a	4:25P LAX	UA 114	FYQM	747	B	0
	5:20P LAX	11:00P O	UA 110	FYBQM	747	D	0
	1:35P	8:32P LAX	AA 10	FYBQM	747	L	0
	12:20a LAX	5:58a O	AA 196	FnYnBQM	D10	S	0
	4:40P	11:30P SFO	UA 188	FYQM	747	D	0
	12:10a SFO	5:55a O	UA 136	FYBQM	D10	S	0
	4:50P	5:52a MSP	NW 22	FYB	747	DS	1
	7:00a MSP	8:05a O	NW 700	FYBQV	72S	B	0
X6	4:50P	5:52a MSP	NW 22	FYB	747	DS	1
	7:00a MSP	8:07a M	NW 800	FYBQV	727	S	0
	6:45P	5:56a DFW	AA 8	FYBQM	747	DS	0
D-27APR	7:01a DFW	9:10a O	AA 242	FYBQM	M80	B	0
X6	8:15P	7:26a DFW	AA 6	FYBQM	747	DS	0
D-27APR	8:20a DFW	10:20a O	BN 30	YBQMV	72S	B	0
	8:15P	7:26a DFW	AA 6	FYBQM	747	DS	0
D-27APR	8:31a DFW	10:35a O	UA 204	FYBQM	727	B	0
X56	8:15P	7:26a DFW	AA 6	FYBQM	747	DS	0
D-27APR	8:30a DFW	10:35a M	ML 356	YM	D9S	S	0
5	8:15P	7:26a DFW	AA 6	FYBQM	747	DS	0
D-27APR	8:30a DFW	10:35a M	ML 356	MK	D9S	S	0
	11:15P	6:05a SFO	UA 210	FYQM	D10	S	0
	7:00a SFO	12:45P O	AA 230	FYBQM	767	B	0
	11:30P	6:10a LAX	AA 2	FYBQM	747	S	0
D-27APR	7:00a LAX	12:40P O	AA 166	FYBQM	D10	B	0
	11:45P	6:50a LAX	UA 6	FYQM	747	S	0
	7:40a LAX	1:25P O	UA 100	FYBQM	D10	B	0
	11:45P	6:50a LAX	UA 6	FYQM	747	S	0
	8:00a LAX	1:35P O	CL 220	Y	D8S	B	0

HOT SPRINGS, ARKANSAS CST HOT
FARES: F 281.00 C 289.00 Y 234.00

Freq.	Leave	Arrive	Flight	Class	Eq	Ml	S
X67	6:50a	8:05a MEM	GM 460	Y	SWM		1
	8:30a MEM	10:00a O	RC 471	CYBMK	D95	B	0
X67	6:50a	8:05a MEM	GM 460	Y	SWM		1
	8:38a MEM	10:00a O	DL 110	FYBMV	72S	S	0
67	9:10a	10:25a MEM	GM 473	Y	SWM		1
	11:10a MEM	12:30P O	RC 475	CYBMK	D95	S	0
X7	10:08a	11:18a MEM	GM 492	Y	SWM		1
	1:02P MEM	2:25P O	DL 1508	FYBMV	D9S	L	0
X67	10:15a	11:10a MEM	GM 482	Y	SWM		0
	1:02P MEM	2:25P O	DL 1508	FYBMV	D9S	L	0
	1:20P	2:40P MEM	GM 407	Y	SWM		1
	3:15P MEM	4:40P O	RC 473	YBMK	DC9		0
X6	6:35P	7:50P MEM	GM 408	Y	SWM		1
	8:15P MEM	9:40P O	RC 483	YBMK	DC9	S	0

HOUSTON, TEXAS CST HOU
H-HOU HAH G-JGP Q-JGQ T-JTC M-JMA P-JPT
FARES: F 200.00-317.00 C 279.00 Y 125.00-264.00
YN 100.00 Q 125.00-155.00 M 189.00
K 189.00
EX: OW 125.00 RT 200.00-349.00

Freq.	Leave	Arrive	Flight	Class	Eq	Ml	S
X7	3:35a I	8:35a O	DL 1418	*	D9S	*	2
			DL1418 FnYnMVBn-ATL-FYBMV				
			DL1418 * MEALS SB/B				
	7:04a I	9:21a O	AA 324	FYBQM	72S	B	0
X7	7:30a H	9:41a O	RC 501	CYBMK	D9S	B	0
	7:51a H	12:45P O	AA 522	FYBQM	727	*	2
			AA 522 * MEALS SL/S				
	7:55a I	11:51a O	AA 280	FYBQM	72S	SB	1
X7	8:00a H	11:59a O	OZ 614	*	D9S	BS	1
			OZ 614 YnBQMV-STL-YBQMV				
X67	8:20a I	10:35a O	CO 154	FY	D9S	B	0
67	8:20a I	10:35a O	CO 154	FYn	D9S	B	0
	8:30a I	10:47a O	UA 316	FYBQM	72S	B	0
	10:25a I	12:38P O	AA 286	FYBQM	72S	L	0
	11:20a H	1:34P O	RC 505	CYBMK	D9S	L	0
X67	11:20a I	1:35P O	CO 150	FY	72S	L	0
67	11:20a I	1:35P O	CO 150	FYn	72S	L	0
	1:00P I	3:16P O	AA 276	FYBQM	727	L	0
	1:38P I	3:55P O	UA 322	FYBQM	72S	S	0
	2:15P H	7:00P O	RC 640	CYBM	D9S	S	2
	2:41P I	5:00P O	AA 130	FYBQM	72S	S	0
	5:00P I	7:21P O	AA 442	FYBQM	72S	D	0
X6	5:05P H	7:19P O	RC 507	CYBMK	D9S	D	0
X6	5:20P I	7:28P O	DL 158	FYBMV	72S	D	0
X6	6:35P I	9:00P O	CO 152	FY	D9S	D	0
6	6:35P I	9:00P O	CO 152	FYn	D9S	D	0
	7:00P I	9:15P O	UA 306	FYBQM	727	D	0
X6	9:00P I	11:15P O	CO 158	FYn	D9S	S/	0

CONNECTIONS
FARES: F 317.00-509.00 C 145.00-279.00
Y 169.00-384.00 M 125.00-232.01
Q 125.00-189.00 K 189.00 V 125.00

Freq.	Leave	Arrive	Flight	Class	Eq	Ml	S
	2:45a I	5:20a ATL	EA 496	FnYnBML	D9S		0
	7:05a ATL	7:50a O	DL 952	FYBMV	D8S	B	0
6	2:45a I	5:20a ATL	EA 496	FnYnBML	D9S		0
	8:00a ATL	8:44a O	EA 1650	FYBML	L10	B	0
X67	6:45a H	8:40a STL	CB 040	YBM	727	B	0
	9:25a STL	10:43a O	UA 534	FYBQM	73S	S/	0
X7	6:55a I	7:50a DFW	BN 140	YBQMV	72S	S	0
	8:20a DFW	10:20a O	BN 30	YBQMV	72S	B	0
	6:55a I	7:50a DFW	BN 140	YBQMV	72S	S	0
	10:00a DFW	12:00n O	BN 34	YBQMV	72S	S	0
X7	7:00a I	8:13a MEM	DL 1483	FYBMV	D9S	B	0
	8:38a MEM	10:00a O	DL 110	FYBMV	72S	S	0
7	7:00a I	8:13a MEM	DL 1455	FYBMV	D9S	B	0
	8:38a MEM	10:00a O	DL 110	FYBMV	72S	S	0
	7:00a I	7:57a DFW	AA 490	FYBQM	727	S	0
	8:40a DFW	10:53a O	AA 470	FYBQM	767	B	0
X7	7:11a I	9:02a STL	TW 566	FYBM	72S	B	0
	10:00a STL	11:06a O	TW 152	FYBM	72S	S/	0
X7	7:20a H	8:13a DFW	DL 1724	FYBMV	73S	S	0
	9:30a DFW	11:32a O	DL 1730	FYBMV	73S	B	0
	8:10a H	10:54a ATL	DL 1710	FYBMV	73S	B	0
	11:41a ATL	12:26P O	DL 1132	FYBMV	L10	L	0
	8:25a I	11:09a ATL	EA 556	FYBML	D9S	B	0
	12:06P ATL	12:44P O	EA 954	FYBML	72S	L	0
	8:50a I	9:46a MSY	EA 68	FYBML	727	S/	0
	12:45P MSY	2:49P O	NW 183	FYBQV	727	L	0
	9:20a I	10:18a DFW	AA 395	FYBQM	72S		0
	11:14a DFW	1:15P O	AA 180	FYBQM	727	L	0
X67	10:40a I	11:38a MSY	CO 151	FY	D9S		0
	12:45P MSY	2:49P O	NW 183	FYBQV	727	L	0
7	10:40a I	11:38a MSY	CO 151	FYn	D9S		0
	12:45P MSY	2:49P O	NW 183	FYBQV	727	L	0
	10:40a I	1:38P CLT	PI 12	YBHQ	72S	L	0
	3:15P CLT	4:05P O	PI 342	YBHQ	73S	S	0
	10:50a I	12:38P STL	TW 482	FYBM	72S	L	0
	1:30P STL	2:32P O	TW 594	FYBM	72S	S/	0
	11:15a I	12:28P MEM	DL 1566	FYBMV	D9S	S	0
	1:02P MEM	2:25P O	DL 1508	FYBMV	D9S	L	0
	11:19a H	12:20P DFW	AA 418	FYBQM	727		0
	1:00P DFW	3:13P O	AA 446	FYBQM	767	L	0

CONT. NEXT COLUMN

To CHICAGO, ILL. CST CHI

From HOUSTON, TEXAS-CONT.

Freq.	Leave	Arrive	Flight	Class	Eq	Ml	S
	11:23a I	12:19P DFW	AA 162	FYBQM	72S		0
	1:00P DFW	3:13P O	AA 446	FYBQM	767	L	0
X6	11:45a H	1:40P STL	CB 042	YBMK	727	L	0
	2:20P STL	3:15P O	DL 1555	FYBMV	D9S		0
X6	11:45a H	1:45P STL	OZ 616	YBQMV	D9S	L	0
	3:00P STL	4:00P O	OZ 668	YBQMV	D9S	S	0
	11:45a H	1:45P STL	OZ 616	YBQMV	D9S	L	0
	4:00P STL	4:59P O	OZ 862	YBQMV	D9S	S	0
	11:50a I	2:30P ATL	EA 602	FYBML	72S	L	0
	3:20P ATL	4:08P O	EA 242	FYBML	757	S	0
	12:10P I	1:05P DFW	BN 54	YBQMV	72S		0
	2:00P DFW	4:00P O	BN 38	YBQMV	72S	S	0
	12:33P I	3:12P ATL	DL 706	FYBMV	72S	L	0
	4:05P ATL	4:50P O	DL 806	FYBMV	767	S	0
	1:20P H	2:35P MEM	RC 396	CYBMK	D9S		0
	3:15P MEM	4:40P O	RC 473	YBMK	DC9		0
2	1:40P I	3:32P STL	TW 272	FYBM	72S	S	0
	4:50P STL	5:57P O	TW 770	FCYBM	72S	S/	0
X2	1:40P I	3:32P STL	TW 272	FYBM	72S	S	0
	4:50P STL	5:59P O	TW 770	FCYBM	747		0
X6	2:10P I	3:05P DFW	BN 166	YBQMV	72S		0
	4:10P DFW	6:10P O	BN 42	YBQMV	72S	D	0
	2:15P I	3:14P DFW	AA 61	FYBQM	72S		0
	4:00P DFW	6:07P O	AA 310	FYBQM	72S	D	0
	2:15P H	4:54P ATL	DL 348	FYBMV	72S	S	0
	5:36P ATL	6:20P O	DL 756	FYBMV	72S	D	0
	3:00P I	5:10P MSY	OQ 972	Y	EMB		2
	6:05P MSY	8:07P O	NW 185	FYBQV	727	D	0
1234	3:05P I	4:49P MCI	CO 163	FY	D9S	S/	0
D-14APR	5:30P MCI	6:40P M	ML 78	QM	DC9	S	0
X6	3:05P I	4:49P MCI	CO 163	FY	D9S	S/	0
E-15APR	5:30P MCI	6:40P M	ML 198	YM	DC9	S	0
57	3:05P I	4:49P MCI	CO 163	FY	D9S	S/	0
D-14APR	5:30P MCI	6:40P M	ML 78	MK	DC9	S	0
	3:20P I	6:00P ATL	EA 564	FYBML	72S	S/	0
	6:55P ATL	7:39P O	EA 244	FYBML	72S	D	0
X6	3:25P H	4:27P DFW	AA 458	FYBQM	727		0
	5:13P DFW	7:22P O	AA 564	FYBQM	767	D	0
	3:30P I	4:30P DFW	AA 444	FYBQM	72S		0
	5:13P DFW	7:22P O	AA 564	FYBQM	767	D	0
	4:10P I	5:23P MEM	DL 1542	FYBMV	D9S		0
	6:20P MEM	7:43P O	DL 990	FYBMV	D8S	D	0
X6	4:15P H	5:08P DFW	DL 1709	FYBMV	73S		0
	5:55P DFW	7:50P O	DL 553	FYBMV	72S	D	0
X6	4:55P H	6:50P STL	CB 044	YBMK	727	D	0
	8:00P STL	8:59P O	OZ 874	YnBQMV	D9S	S	0
	5:00P I	5:55P DFW	BN 86	YBQMV	72S		0
	6:40P DFW	8:40P O	BN 46	YBQMV	72S	D	0
	5:00P I	7:00P STL	TW 526	FYBM	72S	D	0
	8:00P STL	9:02P O	TW 398	FYBM	72S	S/	0
	5:25P H	8:10P ATL	DL 1750	FYBMV	73S	D	0
	9:15P ATL	9:59P O	DL 472	FnYnMVBn	72S		0
	5:45P H	7:30P STL	OZ 612	YBQMV	DC9	D	0
	8:00P STL	8:59P O	OZ 874	YnBQMV	D9S	S	0
	5:50P I	8:29P ATL	DL 218	FYBMV	72S	D	0
	9:15P ATL	9:59P O	DL 472	FnYnMVBn	72S		0
	6:06P H	7:08P DFW	AA 412	FnYnBQM	727		0
	8:05P DFW	10:10P O	AA 292	FYBQM	72S	S	0
	6:07P I	7:06P DFW	AA 544	FnYnBQM	72S		0
	8:05P DFW	10:10P O	AA 292	FYBQM	72S	S	0
	6:45P I	9:25P ATL	EA 336	FYBML	727	D	0
	10:10P ATL	11:00P O	EA 248	FnYnBML	D9S		0
X6	7:35P I	8:25P DFW	DL 578	FnYnMVBn	72S		0
	9:19P DFW	11:15P O	DL 1572	FnYnMVBn	D9S		0
	8:35P I	11:14P ATL	DL 374	FnYnMVBn	72S		0
	12:02a ATL	12:45a O	DL 1092	FnYnMVBn	L10		0

HUNTINGTON, W. VA. EST HTS
FARES: F 170.00 Y 142.00-180.00

Freq.	Leave	Arrive	Flight	Class	Eq	Ml	S
	6:50a	7:37a CLT	PI 305	YBHQ	73S		0
	8:55a CLT	9:45a O	PI 273	YBHQ	73S	B	0
X7	7:00a	7:41a PIT	AL 436	YBMK	B11		0
	8:20a PIT	8:44a O	AL 69	YBM	D9S	B	0
X67	7:10a	7:47a CVG	OH 590	YH	EMB		0
	8:17a CVG	8:15a O	DL 286	FYBMV	72S	S	0
7	10:15a	10:38a CRW	AL 540	YBM	D9S		0
	11:35a CRW	11:53a O	PI 45	YBHQ	72S	S	0
X67	11:15a	11:54a PIT	AL 402	YBMK	D9S		0
	12:35P PIT	12:59P O	AL 99	YBM	D9S	S	0
X7	11:15a	11:54a PIT	AL 402	YBMK	D9S		0
	1:40P PIT	1:59P O	AL 387	YBM	D9S	S	0
	1:50P	2:37P CLT	PI 386	YBHQ	73S		0
	3:15P CLT	4:05P O	PI 342	YBHQ	73S	S	0
X67	4:25P	5:00P CVG	OH 253	YH	EMB		0
	5:30P CVG	5:28P O	DL 134	FYBMV	72S		0
X67	6:25P	7:05P CVG	OH 431	YH	EMB		0
	7:57P CVG	8:00P O	DL 656	FYBMV	72S		0
7	6:50P	7:30P CVG	OH 700	YH	EMB		0
	7:57P CVG	8:00P O	DL 656	FYBMV	72S		0

HUNTSVILLE/DECATUR, ALA. CST HSV
FARES: F 248.00 Y 187.00
EX: 369.00

Freq.	Leave	Arrive	Flight	Class	Eq	Ml	S
	6:45a	10:09a O	EA 240	FYBML	D9S	B/S	1

CONNECTIONS
FARES: F 248.00-371.00 C 202.00 Y 187.00-222.00

Freq.	Leave	Arrive	Flight	Class	Eq	Ml	S
X7	7:10a	7:49a MEM	RC 512	YBM	DC9		0
	8:30a MEM	10:00a O	RC 471	CYBMK	D95	B	0
X7	7:10a	7:49a MEM	RC 512	YBM	DC9		0
	8:38a MEM	10:00a O	DL 110	FYBMV	72S	S	0
	7:39a	10:18a DFW	AA 531	FYBQM	72S	B	1
	11:14a DFW	1:15P O	AA 180	FYBQM	727	L	0
7	9:15a	11:04a ATL	EV 681	YB	EMB		0
	11:41a ATL	12:26P O	DL 1132	FYBMV	L10	L	0
X7	10:00a	10:39a MEM	RC 180	YBM	DC9		0
	11:10a MEM	12:30P O	RC 475	CYBMK	D95	S	0
	10:45a	12:24P ATL	RC 240	YBM	DC9		0
	12:55P ATL	1:35P O	RC 275	CYBMK	D9S	L	0
	11:50a	1:44P ATL	EV 623	YB	EMB		0
	2:26P ATL	3:10P O	DL 759	FYBMV	72S		0
	12:32P	3:13P DFW	AA 457	FYBQM	727	L	1
	4:00P DFW	6:07P O	AA 310	FYBQM	72S	D	0
	2:15P	2:43P BNA	RC 236	YBMK	DC9		0
	3:50P BNA	5:05P O	RC 395	CYBMK	D9S		0
	2:59P	5:48P DFW	AA 193	FYBQM	72S	S	1
	6:35P DFW	8:42P O	AA 476	FYBQM	M80	D	0
X6	3:02P	4:40P ATL	RC 577	YBM	DC9		0
	5:10P ATL	5:50P O	RC 273	CYBMK	D9S	D	0
	4:15P	5:50P ATL	RC 241	YBM	DC9		0
	6:55P ATL	7:39P O	EA 244	FYBML	72S	D	0
X67	4:50P	6:45P ATL	EV 625	YB	EMB		0
	7:45P ATL	8:30P O	DL 920	FYBMV	D8S	S	0
X6	7:00P	7:39P MEM	RC 188	YBMK	DC9		0
	8:15P MEM	9:40P O	RC 483	YBMK	DC9	S	0

HUTCHINSON, KAN. CST HUT
FARES: F 351.99-365.00

Freq.	Leave	Arrive	Flight	Class	Eq	Ml	S
X7	6:29a	6:46a ICT	ZV 820	YBK	SWM		0
	9:05a ICT	10:44a O	UA 758	FYBQM	73S	S	0
X7	6:29a	8:37a MCI	ZV 820	YBK	SWM		2
	9:35a MCI	10:48a O	UA 772	FYBQM	73S		0
X67	12:04P	1:52P MCI	ZV 402	YBK	SWM		1
	2:45P MCI	3:56P O	UA 640	FYBQM	72S		0
7	12:04P	12:30P ICT	ZV 722	YBK	SWM		0
	1:20P ICT	4:04P O	UA 398	FYBQM	73S		1
X67	12:04P	1:52P MCI	ZV 402	YBK	SWM		1
E-15APR	3:30P MCI	4:40P M	ML 150	MK	DC9	S	0
7	12:04P	2:12P MCI	ZV 722	YBK	SWM		2
E-15APR	3:30P MCI	4:40P M	ML 150	MK	DC9	S	0
7	12:04P	2:12P MCI	ZV 722	YBK	SWM		2
	5:32P MCI	6:51P O	TW 454	FYBM	72S	S	0

HYANNIS, MASSACHUSETTS EST HYA
FARES: F 290.00-395.00 Y 267.00-281.01 M 189.00
K 239.00

Freq.	Leave	Arrive	Flight	Class	Eq	Ml	S
1	6:00a	6:40a BOS	PT 720	YBQM	EMB		0
	7:40a BOS	9:05a O	UA 955	FYBQM	D10	B	0

CONT. NEXT PAGE

CAN YOU READ AN AIRLINE LEDGER?

1. What airlines fly to Chicago from Houston?
   - a. _____ American (AA)
   - b. _____ Braniff (BN)
   - c. _____ Continental (CO)
   - d. _____ Delta (DL)
   - e. _____ Eastern (EA)

2. How many direct flights are scheduled daily to Chicago from Hartford, Conn.? _____

3. How many UA (United) flights are scheduled daily to Chicago from Houston? _____

4. Tell the flight number to Chicago:
   - a. _____ from Hartford LV 7:10 am/AR 8:22 am
   - b. _____ from Honolulu LV 5:40 pm/AR 5:35 pm
   - c. _____ from Harrisburg LV 1:50 pm/AR 3:34 pm

5. On these above flights, are meals served? If yes, what kind?
   - a. _____ to Chicago from Hartford
   - b. _____ to Chicago from Honolulu
   - c. _____ to Chicago from Harrisburg

Answers: 1. a) yes b) no c) yes d) yes e) no; 2. 8; 3. three; 4. a) AA 515 b) UA 2 c) AL 169; 5. a) breakfast– B b) dinner/snack– D/S c) no

MARCH 15, 1985/APRIL 27, 1985
AIR ATLANTA SCHEDULE
FOR RESERVATIONS AND INFORMATION, CALL YOUR TRAVEL AGENT OR

Atlanta	(404) 530-2525	Miami	(305) 371-6534
Memphis	(901) 521-1365	New York	(212) 517-8216

Or, 1-800-241-5408

ATLANTA TO MEMPHIS

Flight #	Days	Departure	Arrival	Meals
991	X67	9:25A	9:29A	B
975	X67	1:05P	1:15P	S
971	X67	3:30P	3:40P	S
979	Daily	6:45P	6:55P	D

ATLANTA TO MIAMI

Flight #	Days	Departure	Arrival	Meals
990	X67	9:45A	11:30A	B
992	67 Only	11:20A	1:05P	L
994	X6	1:30P	3:15P	L
996	X67	4:05P	5:50P	S
998	X6	6:55P	8:45P	D

ATLANTA TO NEW YORK (JFK)

Flight #	Days	Departure	Arrival	Meals
972	X67	7:30A	9:30A	B
970	X67	10:30A	12:30P	L
976	Daily	1:15P	3:14P	L
974	Daily	4:30P	6:45P	D
978	X67	6:30P	8:30P	D

MEMPHIS TO ATLANTA

Flight #	Days	Departure	Arrival	Meals
980	X67	6:30A	8:30A	B
992	67 Only	9:00A	11:00A	S
976	X67	10:45A	12:50P	S
974	X67	1:40P	3:45P	S
978	X67	4:00P	6:05P	S

MEMPHIS TO MIAMI

Flight #	Days	Departure	Arrival	Meals
994	X67	10:45A	3:15P	SL
992	67 Only	9:00A	1:05P	SL
996	X67	1:40P	5:50P	S
998	X67	4:00P	8:45P	SD

MEMPHIS TO NEW YORK (JFK)

Flight #	Days	Departure	Arrival	Meals
976	67 Only	9:00A	3:14P	SL
976	X67	10:45A	3:14P	SL
974	X67	1:40P	6:45P	SD
978	X67	4:00P	8:30P	SD

MIAMI TO ATLANTA

Flight #	Days	Departure	Arrival	Meals
991	X67	7:25A	9:10A	B
995	Daily	1:15P	3:05P	L
997	X6	4:30P	6:20P	D
999	X7	6:30P	8:20P	D

MIAMI TO MEMPHIS

Flight #	Days	Departure	Arrival	Meals
991	X67	7:25A	9:29A	B
995	X67	1:15P	3:40P	LS
997	X6	4:30P	6:55P	DD

NEW YORK (JFK) TO ATLANTA

Flight #	Days	Departure	Arrival	Meals
973	X1	8:15A	10:25A	B
975	X67	10:30A	12:40P	L
971	X67	1:00P	3:10P	L
979	Daily	3:45P	6:05P	S
977	X6	7:15P	9:35P	D

NEW YORK (JFK) TO MEMPHIS

Flight #	Days	Departure	Arrival	Meals
975	X67	10:30A	1:15P	LS
971	X67	1:00P	3:40P	LS
979	Daily	3:45P	6:55P	SD

WINTER SPECIAL 12/15/84 — 3/30/85 ★
ATLANTA TO GUNNISON, COLO.

Flight #	Days	Departure	Arrival	Meals
901	Sat. Only	10:00A	11:30A	B

GUNNISON, COLO. TO ATLANTA

Flight #	Days	Departure	Arrival	Meals
902	Sat. Only	12:15P	6:15P	L

Legend
X6 = Except Saturday
X7 = Except Sunday   X1 = Except Monday
X67 = Except Saturday, Sunday   67 Only = Sat./Sun. Only
★ = Guest passes not valid on these special flights

Reprinted with permission from Air Atlanta.

AIR TRAFFIC PATTERNS

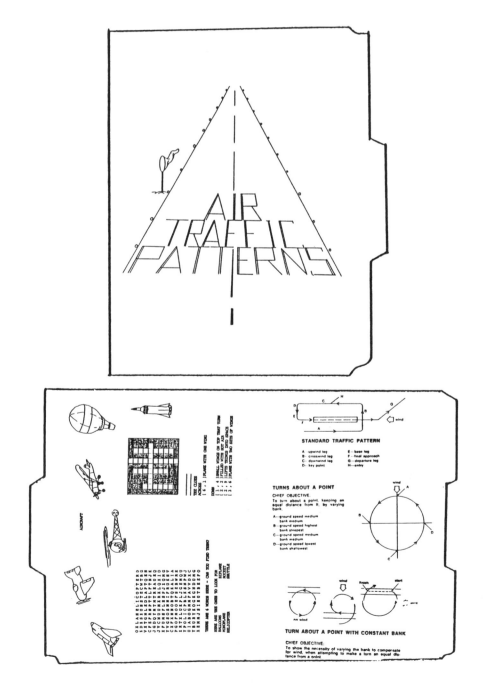

NAME: Air Traffic Patterns/Basic Air Maneuvers

SKILL: Geometric patterns and shapes

PROCEDURE: Study the diagrams. Cut out a small plane or use a model plane to demonstrate the traffic patterns. Play the word search and crossword puzzle.

VARIATIONS:
1. See if you can find out what factors determine when to use the different traffic patterns.
2. Draw and color the six aircraft described on the worksheet.

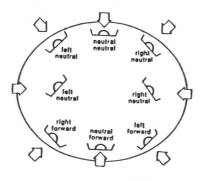

ARROWS INDICATE
WIND DIRECTION

Taxi Diagram
(tricycle gear aircraft)

Standard Traffic Pattern

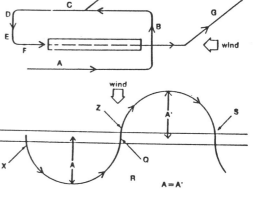

A — Upwind leg
B — Crosswind leg
C — Downwind Leg
D — Key Point

E – Base leg
F — Final Approach
G — Departure leg
H — Entry

S-Turns Across a Road

Chief Objective: To fly two perfect half circles on opposite sides of the road, by varying bank to compensate for the wind.

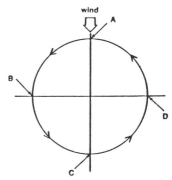

X & S — ground speed highest, bank steepest
R & Z — ground speed lowest, shallowest bank
Q — wings level

Turns About a Point

CHIEF OBJECTIVE: To turn about a point, keeping an equal distance from it, by varying bank.

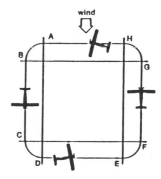

A — ground speed medium, bank medium
B — ground speed highest, bank steepest
C — ground speed medium, bank medium
D — ground speed lowest, bank shallowest

Rectangular Course

A — Medium
B — Steep
C — Steep
D — Medium

E — Medium
F — Shallow
G — Shallow
H — Medium

CHIEF OBJECTIVE: To maintain an equal distance around a rectangle, through a combination of crabbing and bank variation to compensate for the wind.

Turn About a Point with Constant Bank

CHIEF OBJECTIVE: To show the necessity of varying the bank to compensate for the wind, when attempting to make a turn an equal distance from a point.

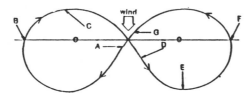

77

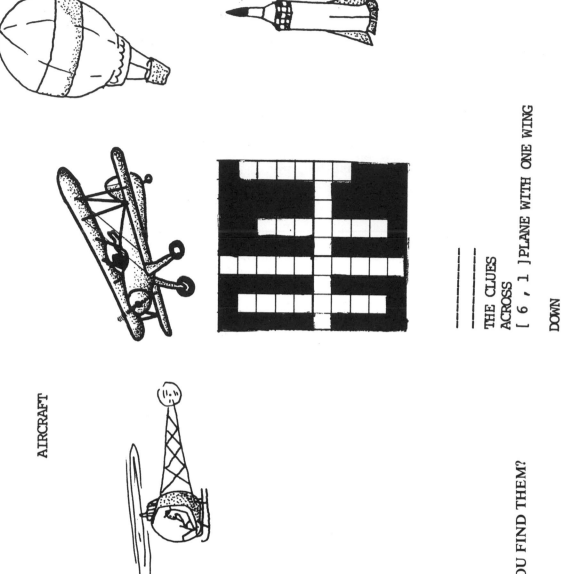

AIRCRAFT

THE CLUES
ACROSS
[6 , 1] PLANE WITH ONE WING

DOWN
[1 , 4] SMALL WINGS ON TOP THAT TURN
[2 , 2] FILLED WITH HOT AIR
[2 , 9] LIFTS THINGS INTO SPACE
[3 , 6] PLANE WITH TWO SETS OF WINGS

O X L A H E L I C O P T E R D
Z D Z S Z U H S T E C W T B G
V X T G N L E D L A P T Q E N
C F E F R N P T T L F T N J K
J G D Q Y M T M T R E A U E Q
J V C N U M X R X L Q X I O O
X F U M H F L H S P J Z Q M G
Y V G S O S N N I F R U K H V
C X F X N N O B T Q K A L I A
P G O F M O O E F S E B W R K
Z R T X L U K P L J G E M Q U
Y P O L I C P B L K S N P W J
U I A B O G M M F A V V D X C
Y B U R Y K O G K G N D U M G
X E M M J S I H U T B E E F O

THERE ARE 6 WORDS HERE — CAN YOU FIND THEM?

HERE ARE THE ONES TO LOOK FOR:
Balloons   Biplane
Monoplane   Rocket
Helicopter   Shuttle

78

```
. . . . H E L I C O P T E R .
. . . . . . . E . . . . . E .
. . . . . . L . . . E .
. . . . . T . . N . .
. . . . T . . . A . . .
. . . U . . . L . . . .
. . M H . . . S P . . . . .
. . S O . . N I . . . . . .
. . . . N O B T . . . . . .
. . . . O O E . . . . . .
. . . L . K P . . .
. . L . C . . L . . . .
. A . O . . . A . . . .
. B . R . . . . N . . .
. . . . . . . . . E . . .
```


THE ANSWERS
ACROSS
[6 , 1]MONOPLANE

DOWN
[1 , 4]HELICOPTER
[2 , 2]BALLOONS
[2 , 9]ROCKET
[3 , 6]BIPLANE

79

79

NAME: Aviation Careers

SKILL: Career Awareness

PROCEDURE: Draw a gameboard on the inside of the folder. Cut out the career cards, bend on the dotted lines and tape the open edge. Mix up the cards. Throw a die and move the correct number of spaces . You must answer the role of the career before you are allowed to stay on the space. Continue procedure until there is a winner.

VARIATIONS: 1. Cut the cards on the dotted line and the solid lines. See if you can match the person with his career description
2. Read the FAA career awareness series to find out more about a particular job career that you may be interested in.

FLIGHT SERVICE SPECIALIST	Talks to the private pilot and gives weather and flying information, and helps plan the flight.
FLIGHT DISPATCHER	Helps the pilot plan the flight.
GROUND CREW	They put fuel in the plane. They also tell the pilot where to park the plane and help the passengers get off the plane.
GROUND CONTROLLER	Controls the airplane on the ground. Tells the pilot where to taxi.
TOWER AIR TRAFFIC CONTROLLER	Gives the pilot permission to take off and land; communicates on a two-way radio; controls air traffic.
AIR ROUTE TRAFFIC CONTROLLER	Watches the plane on the radar screen and tells the pilot where to fly along the airways.
PORTER	Helps the passengers carry their baggage into the terminal.
AIRLINE WORKER AT THE TICKET COUNTER	Sells you your ticket, gives you a boarding pass, checks your baggage.
AIRCRAFT ENGINE MECHANICS	They check the engine and instruments to see if they are working properly on the plane.

CARGO LOADERS	They put baggage, mail, and supplies on board the plane.
FLIGHT ATTENDANTS	They work in the passenger cabin making sure everyone is comfortable. They serve drinks and meals. They help passengers find their seats and fasten their seatbelts and demonstrate emergency procedures.
COPILOT	Sits in the cockpit with the pilot and helps fly the plane.
FLIGHT ENGINEER	Checks the equipment and makes sure it operates properly during flights. Sits in the cockpit behind the pilots.
PILOT	Sits in the cockpit and flys the plane. This person studies weather maps and files flight plans.

AIRCRAFT

Collage

NAME: Aircraft Collage

SKILL: Visual Awareness

PROCEDURE: Study the collage with the aircraft. Find all the things that fly and write your answers on the laminated folder. Show your answers to your teacher before erasing them.

VARIATIONS: 1. Make your own aerospace collage with magazine pictures and/or draw your own collages.
2. Using a variety of media, create your own collage.

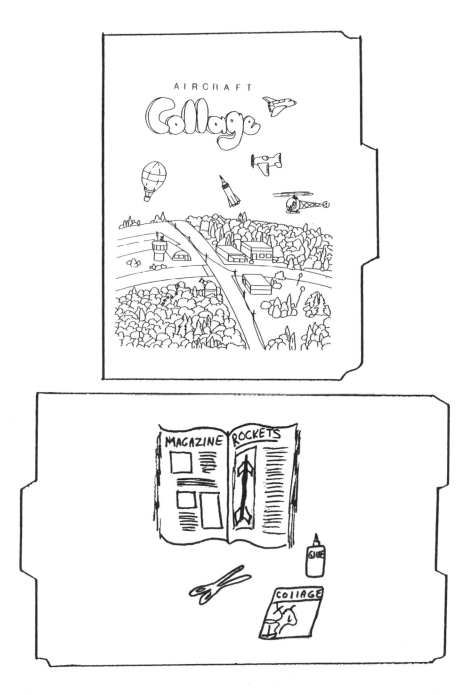

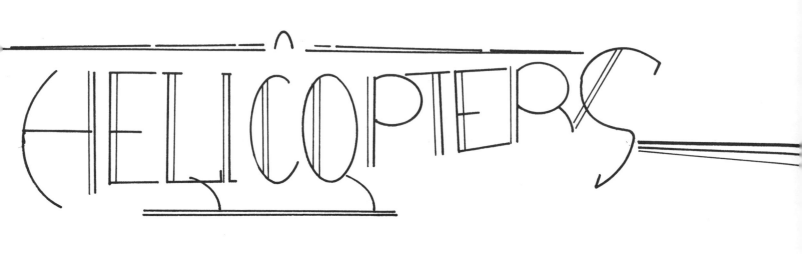

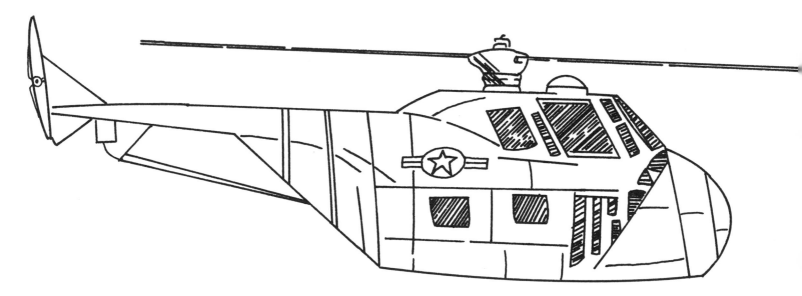

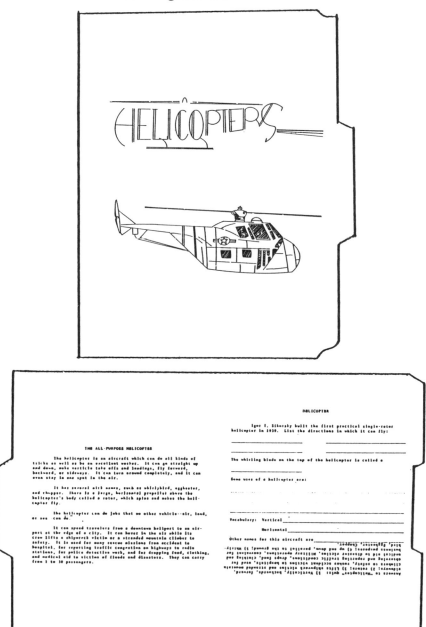

NAME: Helicopters

SKILL: Reading Comprehension

PROCEDURE: 1. Read the story about the helicopter and answer the questions. Check your answers with the answer key, then erase your answers.
2. Make the paper helicopter and observe how it flies.

VARIATIONS: 1. Read more about the uses of the helicopter. Write a paper on the numerous uses of helicopters.
2. Find out more about the life and discoveries of Leonardo da Vinci and Igor I. Sikorsky.
3. Visit a local airport to see a helicopter — compare it to a fixed-wing aircraft.

THE ALL-PURPOSE HELICOPTER

The helicopter is an aircraft which can do all kinds of tricks as well as be an excellent worker. It can go straight up and down, make vertical take offs and landings, fly forward, backward, or sideways. It can turn around completely, and it can even stay in one spot in the air.

It has several slick names, such as whirlybird, eggbeater, and chopper. There is a large, horizontal propellor above the helicopter's body called a rotor, which spins and makes the helicopter fly.

The helicopter can do jobs that no other vehicle--air, land, or sea--can do.

It can speed travelers from a downtown heliport to an airport at the edge of a city. It can hover in the air while its crew lifts a shipwreck victim or a stranded mountain climber to safety. It is used for many rescue missions from accident to hospital, for reporting traffic congestion on highways to radio stations, for police detective work, and for dropping food, clothing, and medical aid to victims of floods and disasters. They can carry from 1 to 30 passengers.

HELICOPTER

Igor I. Sikorsky built the first practical single-rotor helicopter in 1939. List the directions in which it can fly:

_____

_____

The whirling blade on the top of the helicopter is called a

_____

Some uses of a helicopter are:

_____

_____

Vocabulary: Vertical_____

Horizontal_____

Other names for this aircraft are_____

Answers: "Helicopter": 1) Vertically, backward, forward, sideways, around. 2) Rotor. 3) Little hospitals if flying accidents and medical aid to victims. Vocabulary: Vertical--up and down, Horizontal--sideways. Other names for this aircraft are whirlybird, eggbeater and chopper.

THE ALL-PURPOSE HELICOPTER

The idea of the helicopter was thought of hundreds of years ago. The Chinese mady toys which were flying tops with rotors made of feathers. Leonardo da Vinci, (1452 – 1519) a great Italian artist and scientist, drew sketches of a similar flying machine.

Igor I. Sikorsky built the first practical single-rotor helicopter in 1939 and flew it in 1940. He was a Russian engineer who came to this country in 1919. The United States government used many Sikorsky helicopters for its Armed Forces.

The helicopter is an aircraft which can do all kinds of tricks as well as be an excellent worker. It can go straight up and down, make vertical take offs and landings, fly forward, backward, or sideways. It can turn around completely, and it can even stay in one spot in the air.

It has several nick-names, such as whirlybird, eggbeater, and chopper. There is a large, horizontal propeller above the helicopter's body called a rotor, which spins and makes the helicopter fly.

The helicopter can do jobs that no other vehicle — air, land or sea — can do.

It can speed travellers from a downtown heliport to an airport at the edge of a city. It can hover in the air while its crew lifts a shipwreck victim or a stranded mountain climber to safety. It is used for many rescue missions from accident to hospital, for reporting traffic congestion on highways to radio stations, for police detective work, and for dropping food, clothing, and medical aid to victims of floods and disasters. They can carry from 1 to 30 passengers.

HELICOPTER

Igor Sikorsky built the first practical single-rotor helicopters in 1939. List the directions in which it can fly:

_____   _____

_____   _____

The whirling blade on the top of the helicopter is called a _____

Some uses of a helicopter are:

_____

_____

_____

Vocabulary: Vertical _____

Horizontal _____

Other names for this aircraft are _____

_____

Answers to "Helicopter" quiz: 1) Vertically, backwards, forward, sideways; 2) rotor 3) Lifts shipwreck victims and stranded mountain climbers to safety, rushes accident victims to hospitals, used for observing and reporting traffic conditions, drops food, clothing and medical aid to disaster victims, Military operations, convenient for business purposes; 4) up and down, parallel to the ground; 5) Whirlybird, Eggbeater, Chopper.

Helicopter

1. Cut out a strip of paper 8 x 2 inches (the bottom two inches of a piece of notebook paper works perfectly). Make a cut down the center from the top 4½ inches long and from each side at a point 3 inches from the bottom ⅔ inches long.
2. Fold the left bottom panel evenly over the middle.
3. Fold the right bottom panel evenly over the left.
4. Fold the left bottom corner across the middle, and the right corner the opposite way across the back to make a point.
5. Fold the left top strip forward, the right top strip backward. These are the rotors.
6. Holding the helicopter high above your head, release it. Its spinning flight downward demonstrates "counterrotation," a method of emergency descent for helicopters when the engine fails.

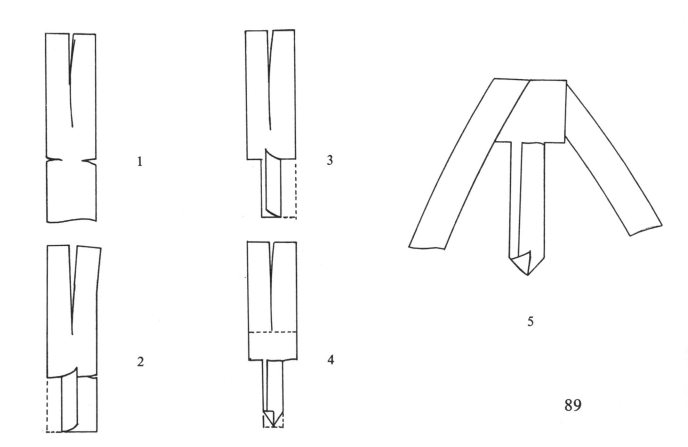

1

2

3

4

5

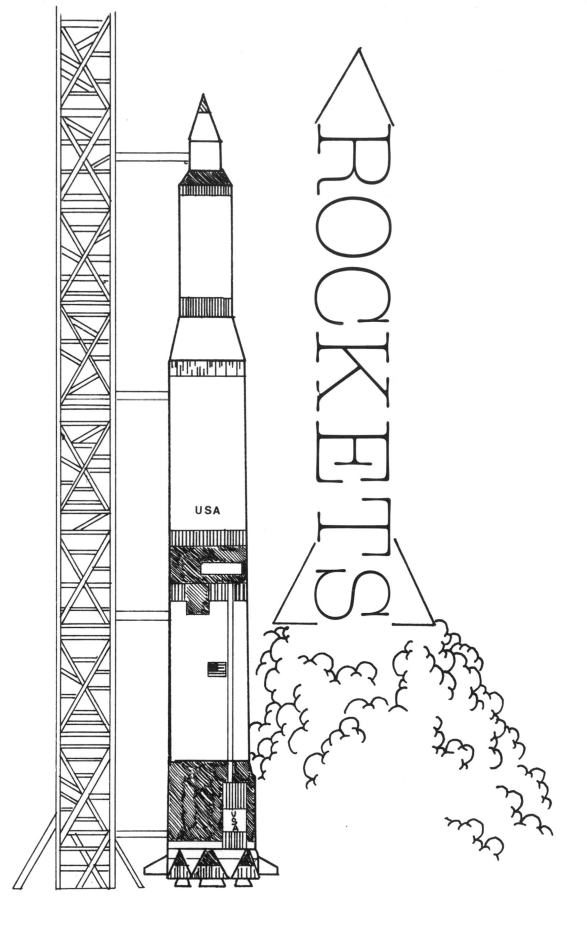

USA

ROCKETS

NAME: Rockets

SKILL: Art and Sequence Activity

PROCEDURE: Cut out the rocket sequence pictures. Attach at the top with a staple. Flip at the bottom and the sequence will appear.

VARIATIONS: 1. Watch for cartoon strips that show sequence relation to aviation/space concepts. Make a flip book with the individual frames.
2. Arrange the pictures in sequential order. Place numerals on the back for self-correcting.

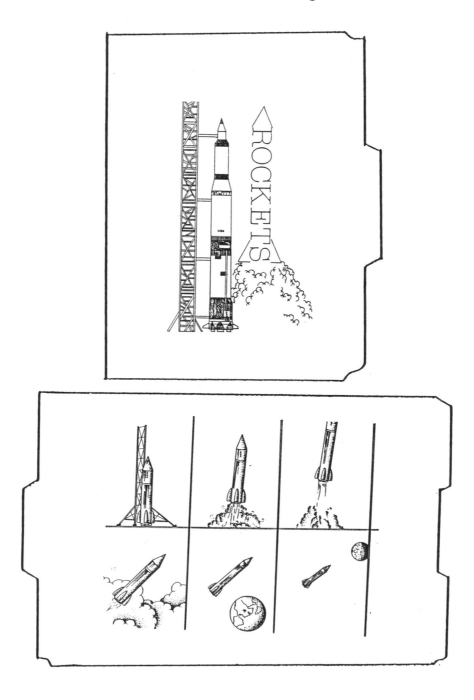

TYPES OF ROCKETS

1. Sky Rocket
2. Propellant charge (solid fuel — a fuel plus a chemical that contains oxygen)
3. Mercury-Redstone (Liquid fuel — liquid gas plus liquid oxygen)
4. Saturn V (#3 stage liquid fuel rocket)

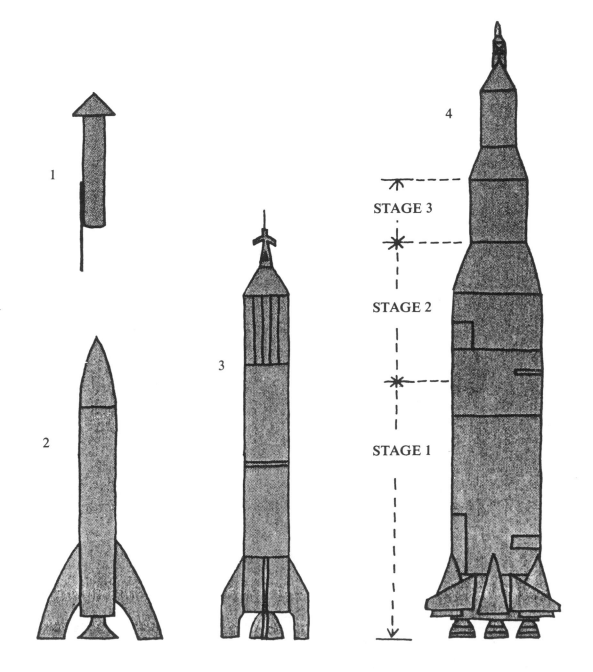

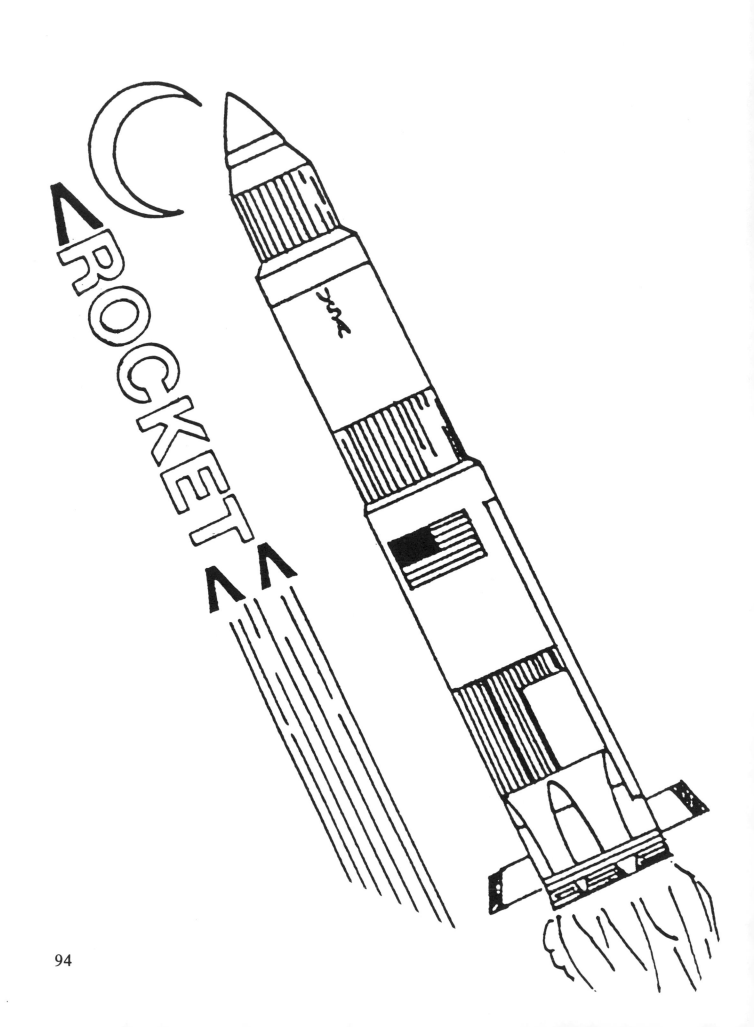

A ROCKET

94

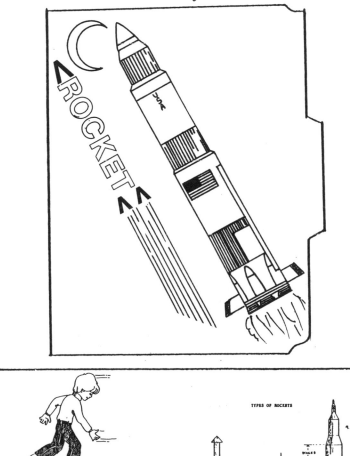

NAME: Rocket!

SKILL: Reading Comprehension and Art

PROCEDURE: Read how a rocket works. Then learn the rocket poems and demonstrate with a cardboard cylinder or paper rocket that you make, following the directions provided.

VARIATIONS: 1. Read more about rockets from the various NASA publications received.
2. Visit the library to find out about ancient Chinese rockets.
3. Find out what Newton's first two laws of motion demonstrated.
4. Research the library to find out about Robert Goddard, the Father of Rocketry.

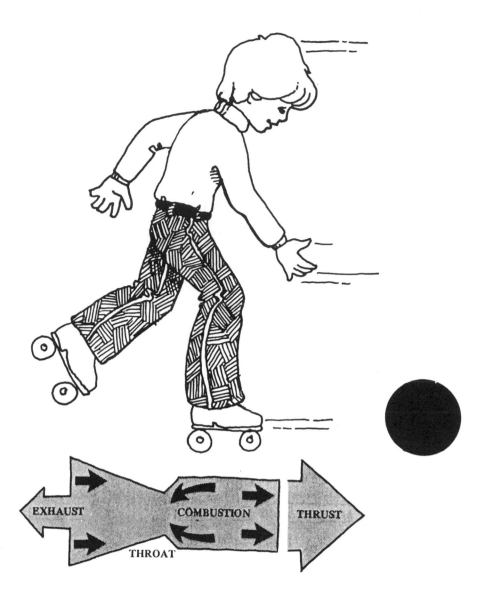

MAKE YOUR OWN ROCKET

The rocket is based on the principle called Newton's Third Law of Motion. It states that for every action there is an equal and opposite reaction. In the picture you can see this illustrated. The boy is standing on roller skates and holding a bowling ball. When he throws the ball there is a reaction force pushing him back in the opposite direction. The simplest rocket you can make is with a toy balloon. The balloon rocket used the principles of two laws of motion. When the balloon is inflated the pressures acting against the wall of the balloon are in balance. When the outlet is opened, gas discharges through the opening and the balloon moves in the opposite directon. This is the same principle of the actual rocket engine.

DID YOU EVER SEE A ROCKET?
(To the Tune of "Did You Ever See A Lassie?")
Did you ever see a rocket, a rocket, a rocket?
Did you ever see a rocket go this way and that?
Go this way and that way, go this way and that way,
Did you ever see a rocket go this way and that?
  (Repeat using other aerospace items)

I'M A ROCKET!

I'm a little rocket
  (Child crouches on heels)
Pointing to the moon
  (Child points upward)

4. . . 3 . . . 2 . . . 1 . . .
  (Said slowly)
Blast off! ZOOM!
  (Springs up and jumps in the air)

96

ROCKETS

MATERIALS: Empty cardboard cylinders
Rolled up paper
Tape
Paint
Any other materials the children can think of . . . let them be creative!

PROCEDURE: Have the children study pictures of rockets and then design their own. They may like to decide where they will blast off to and what they will discover. This activity may also be used to encourage them to write a story about their trip.

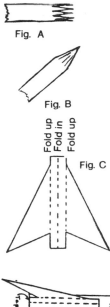

Fig. A

Fig. B

Fold up Fold in Fold up

Fig. C

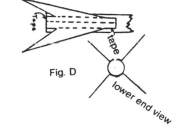

tape

Fig. D

lower end view

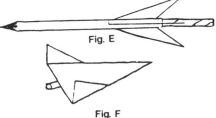

Fig. E

Fig. F

1. Cut a strip of paper about 8" long and 1 to 2" wide.
2. Roll the paper strip around the pencil and tape. The paper tube should slide easily off the pencil but not be too loose.
3. Make several pointed cuts at one end of the tube. See Figure A.
4. Slide the sharpened end of the pencil toward the pointed cuts. Fold the points around the sharpened end of the pencils and tape to form the nose cone. See Figure B.
5. Cut out two sets of fins. Use the pattern in Figure C. Fold the fins on the dashed lines in the manner shown in Figure C.
6. Using two pieces of tape, fix the fins to the opposite end of the tube from the nose cone. Insert the pencil for support in taping. See Figure D.
7. Place the rocket over the soda straw. See Figure E.
8. Select a firing range to launch the rocket such as one side of the classroom or a hallway. A high ceiling room such as an auditorium or gymnasium is best.
9. Launch the rocket by blowing sharply on the straw. Be sure to aim the rocket in the desired direction.

Reprinted with permission from Dr. Ted Colton. 97

Trip to the MOON

98

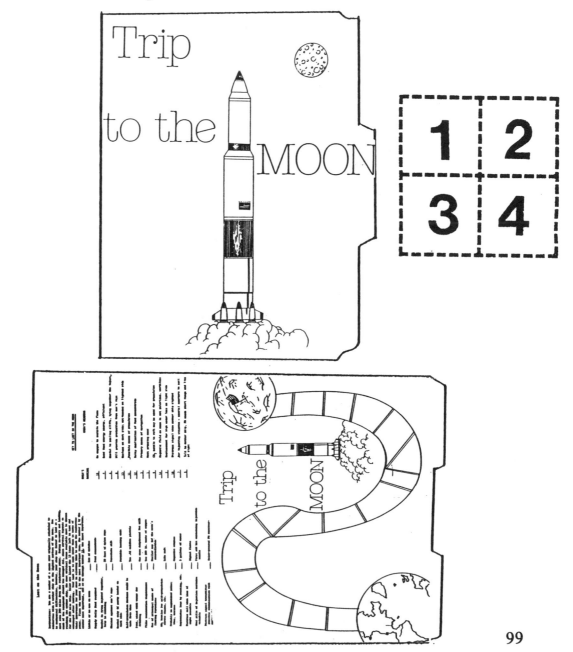

NAME: Trip to the Moon

SKILL: Mathematics

PROCEDURE: 1. Play the board game to see who gets to the moon first. Any skill may be placed on the gameboard. Cut out the numbers and put in a cup. Pick a number and move that number of spaces. You must answer the problem correctly to stay on the space. The first one to the moon wins!

2. See if you can survive on the moon by answering the Moon Survival game.

VARIATIONS: 1. Substitute any skill on the gameboard.

2. Read the NASA publications about the moon and the Apollo.

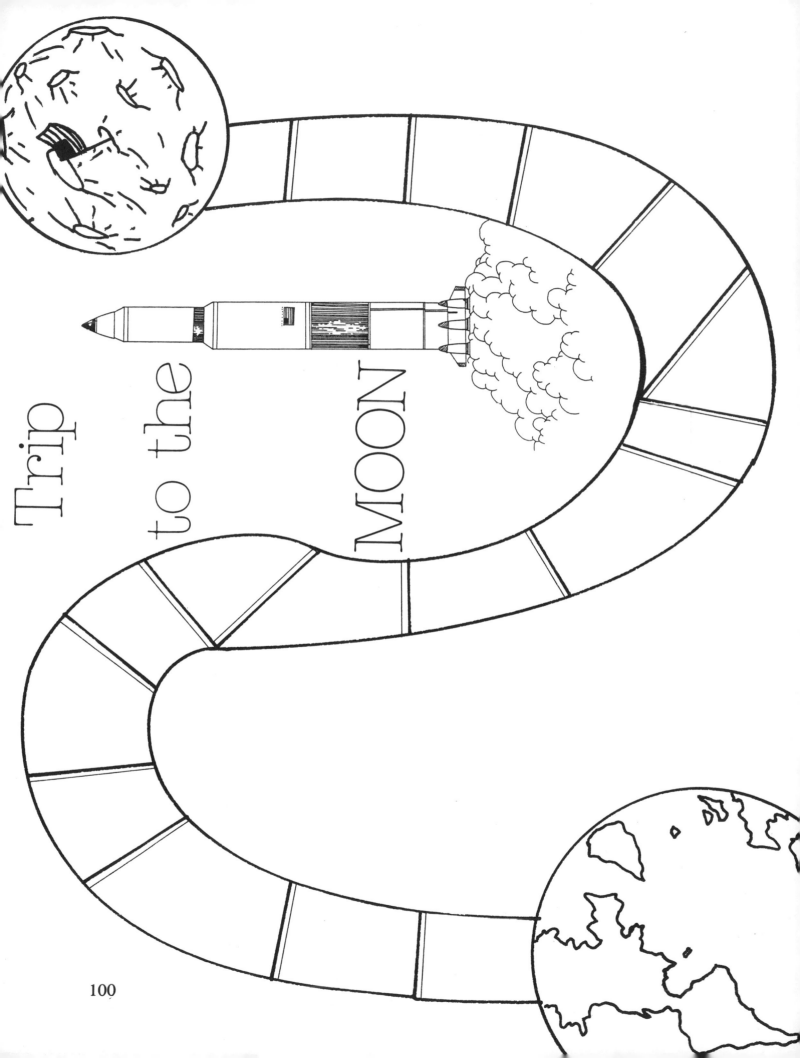

Trip to the MOON

100

MOON SURVIVAL

Instructions: You are a member of a space crew originally scheduled to rendezvous with a mother ship on the lighted surface of the moon. Due to mechanical difficulties, however, your ship was forced to land at a spot some 200 miles from the rendezvous point. During re-entry and landing, much of the equipment aboard was damaged and, since survival depends on reaching the mother ship, the most critical items available must be chosen for the 200 mile trip. Below are listed the 15 items left intact and undamaged after landing. Your task is to rank order them in terms of their importance for your crew in allowing them to reach the rendezvous point. Place the number *1* by the most important item, the number *2* by the second most important, and so on through number *15*, the least important.

_____ Box of matches

_____ Food concentrate

_____ 50 feet of nylon rope

_____ Parachute silk

_____ Portable heating unit

_____ Two .45 calibre pistols

_____ One case dehydrated Pet milk

_____ Two 100-pound tanks of oxygen

_____ Stellar map (of the moon's surface)

_____ Life raft

_____ Magnetic compass

_____ 5 gallons of water

_____ Signal flares

_____ First aid kit containing injection needle

_____ Solar-powered FM receiver-transmitter

KEY

NASA'S RANKING	NASA'S REASONS
15	No oxygen to sustain the flame
4	Good food source, efficient
6	Useful in scaling cliffs, tying together the injured
8	Will provide protection from sun's rays
13	Useless on dark side, not needed on lighted side
11	Possible means of propulsion
12	Bulky duplication of food concentrate
1	Most pressing need
3	Primary means of navigation
9	CO_2 bottle in raft may be used for propulsion
14	Magnetic field on moon is not polarized, worthless
2	Replacement for high water loss on light side
10	Distress signal when mother ship sighted
7	For injecting vitamins — special aperture in suit
5	Talk to mother ship, FM needs short range and line of sight

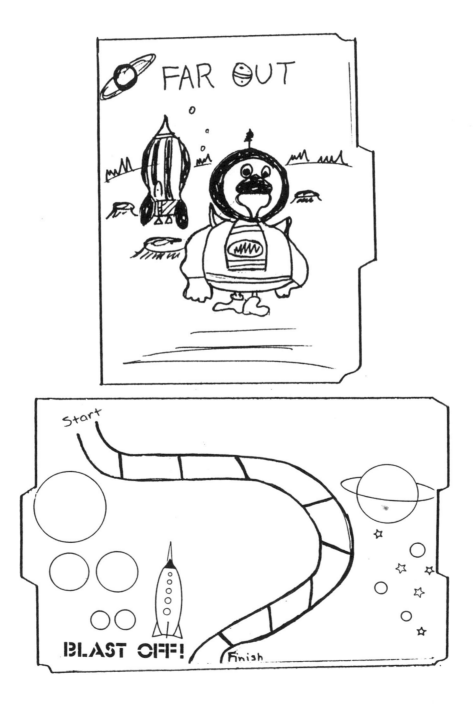

NAME: Far Out!

SKILL: Mathematics

PROCEDURE: Draw the board game with any skill (i.e., math facts) on the squares. Throw a die and move the correct number of spaces. You must say the answer correctly to stay on the space. The opponent then throws the die and repeats the procedure until there is a winner.

VARIATIONS: Substitute any skill (math facts, vocabulary, letter, etc.) on the spaces or change the mathematical problem.

FAR OUT

Start

BLAST OFF!   Finish

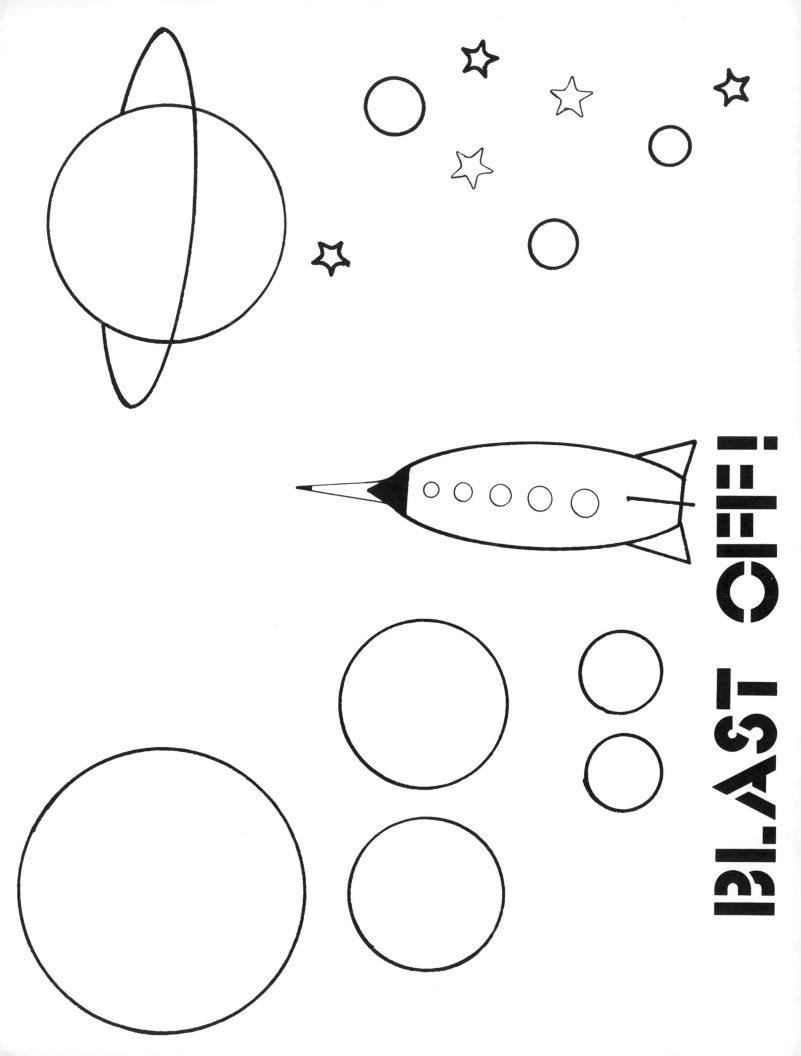

BLAST OFF!

A·S·T·R·O·N·A·U·T

A S T R O N A U T S

NAME: Ten Little Astronauts

SKILL: Poetry and Vocabulary Activity

PROCEDURE: Learn the poems. Glue the astronauts to a file folder and mount to popsicle sticks. Use your astrosticks to act out the poems. Make up your own poems.

VARIATIONS:
1. Make a moon landscape. Use the astrosticks with a lunar landing module for a moonscape model.
2. Read about the Eagle and astronauts Neil Armstrong and Buzz Aldrin, the first men on the moon.
3. Read the NASA publications on the various astronauts and their selections, training and missions.
4. Read about Col. Frederick D. Gregory, space shuttle pilot.

ASTRONAUT POEMS

TWO LITTLE ASTRONAUTS

Two little astronauts are going to the moon.
Two little astronauts hope they'll get there soon.
The first one said, "Oh this is such fun."
The second one said, "We will see the sun."
Then — 10, 9, 8, 7, 6, 5, 4, 3, 2, 1 ZOOM!

TEN LITTLE ASTRONAUTS
(To the Tune of Ten Little Indians)

One little, two little, three little astronauts,
Four little, five little, six little astronauts
Seven little, eight little, nine little astronauts,
Ten little astronauts.

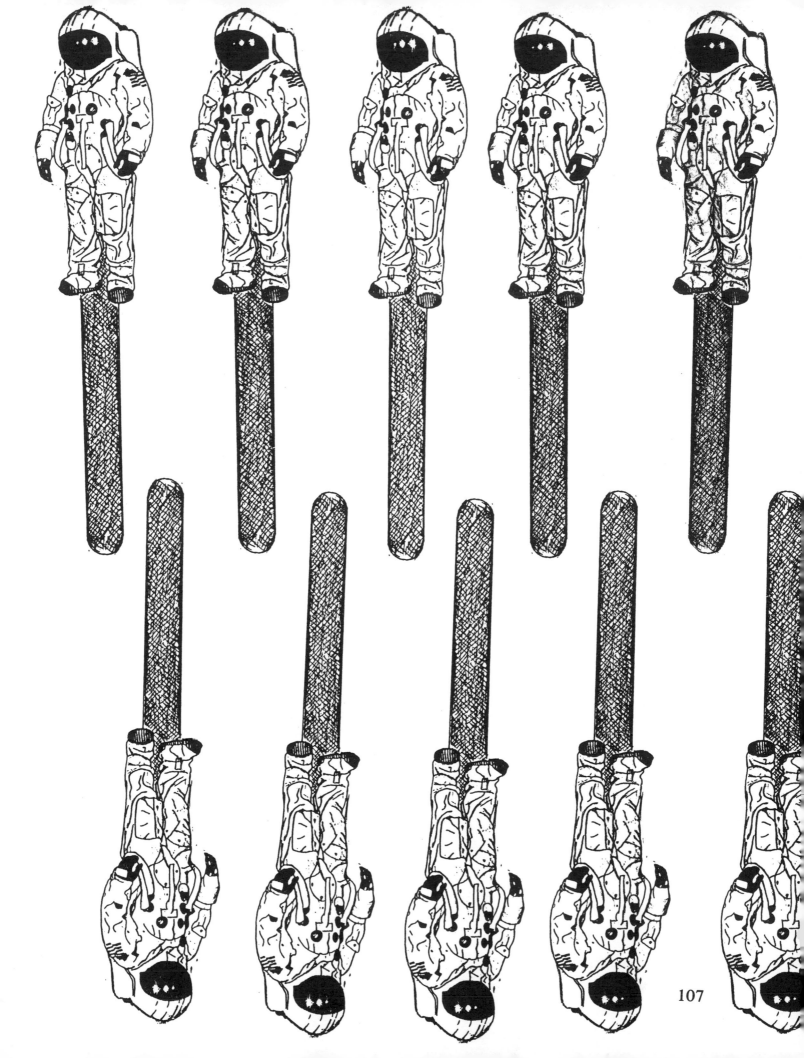

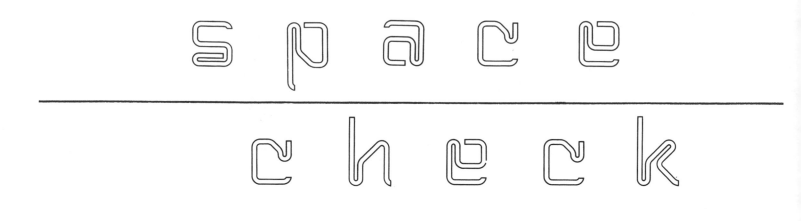

space

check

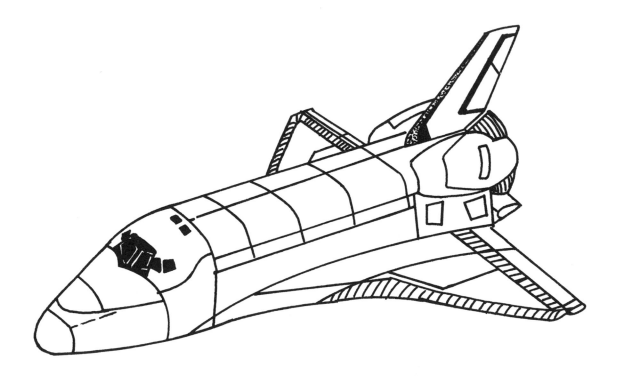

NAME: Space Check

SKILL: Health and Food Awareness

PROCEDURE: Read about the kinds of food that are used in space. Look for freeze dried or dried foods, such as Tang , in the grocery store. Check your physical condition to see how fit you are to be an astronaut.

VARIATIONS: 1. Plan a balanced menu for space flight.
2. Program a physical fitness program for yourself to enhance your fitness for space — include a record of your diet and exercise program.

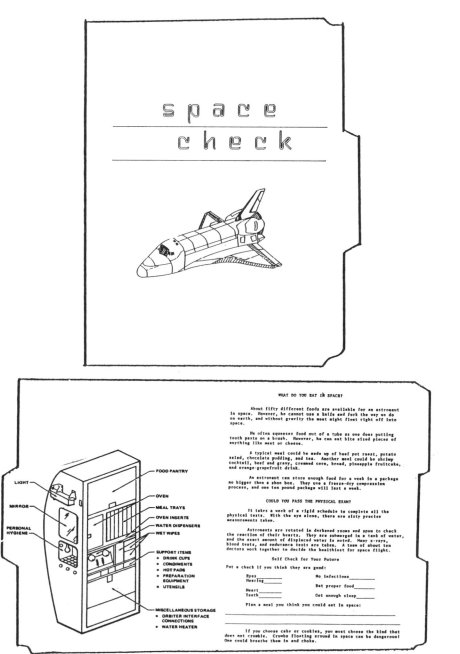

FOOD SYSTEM

The Orbiter is equipped with food, food storage, and food preparation and dining facilities to provide each crewman with three meals per day plus snacks and an additional 96 hours of contingency food. The food supply and food preparation facilities are designed to accommodate flight variations in the number of crewmen and flight durations ranging from two crewmen for 1 day to seven crewmen for 30 days.

The food consists of individually packaged serving portions of dehydrated, thermostabilized, irradiated, intermediate moisture, natural form, and beverage foods. The food system relies heavily on dehydrated food, since water is a byproduct of the fuel cell system onboard the Orbiter. Off-the-shelf thermostabilized cans, flexible pouches, and semirigid plastic containers are used for food packaging.

The menu will be a standard 6-day menu instead of the personal-preference types used in previous programs. The menu will consist of three meals each day plus additional snacks and beverages.

The daily menu will be designed to provide an average energy intake of 3000 calories for each crewmember. The food system also includes a pantry of foods for snacks and beverages between meals and for individual menu changes. The galley, which is located in the cabin working area is modular and can be removed for special missions. In addition to cold and hot water dispensers, it will be equipped with a pantry, an oven, food serving trays, a personal hygiene station, a water heater, and auxiliary equipment storage areas. The oven will be a forced-air convection heater with a maximum temperature of 355 K (82° C of 180° F). There are no provisions for food freezers or refrigerators.

Food preparation activities will be performed by one crewman 30 to 60 minutes before mealtime. The crewman will remove the selected meal from the storage locker, reconstitute those items that are rehydratable, place the foods to be heated into the galley oven, and assemble other food items on the food trays. Meal preparation for a crew of seven can be accomplished by one crewmemember in about a 2 minute period. Utensils and trays are the only items that require cleaning after a meal. Cleaning will be done with sanitized "wet wipes" that contain a quaternary ammonium compound. As on the Skylab flights, the crewmembers will use regular silverware.

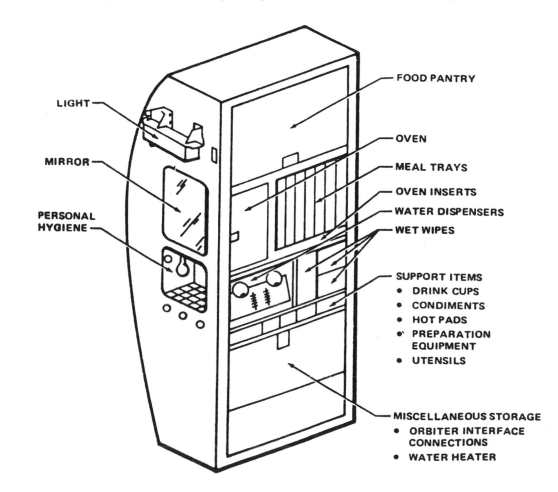

COULD YOU PASS THE PHYSICAL EXAM?

It takes a week of rigid schedule to complete all the physical tests. With the eye alone, there are sixty precise measurements taken.

Astronauts are rotated in darkened rooms and spun to check the reaction of their hearts. They are submerged in a tank of water, and the exact amount of displaced water is noted. Many x-rays, blood tests, and endurance tests are taken. A team of about ten doctors work together to decide the healthiest for space flight.

SELF CHECK FOR YOUR FUTURE

Put a check if you think they are good:

Eyes ————

Hearing ————

Heart ————

Teeth ————

No Infections ————

Eat proper food ————

Get enough sleep ————

WHAT DO YOU EAT IN SPACE?

About fifty different kinds of foods are available for an astronaut in space. However, he cannot use a knife and fork the way we do on earth, since without gravity, the meat might float right off into space.

He often squeezes food out of a tube as one does putting toothpaste on a brush. However, he can eat bite sized pieces of anything like meat or cheese.

A typical meal could be made up of beef pot roast, potato salad, chocolate pudding, and tea. Another meal could be shrimp cocktail, beef and gravy, creamed corn, bread, pineapple fruitcake, and orange-grapefruit drink.

An astronaut can store enough food for one week in a package no bigger than a shoe box. They use a freeze-dry compression process, and one ten-pound package will last a week.

Plan a meal you think you could eat in space.

_____

_____

_____

If you choose cake or cookies, you must choose the kind that does not crumble. Crumbs floating around in space can be dangerous! One could breathe them in and choke.

TYPICAL SPACE SHUTTLE MENU

Dried Apricots (IM)	Macaroni and cheese (R)	Noodles and chicken (R)
Breakfast roll (I, NF)	Peas w/butter sauce (R)	Stewed Tomatoes (T)
Granola with blueberries (R)	Peach ambrosia (R)	Pears (FD)
Vanilla instant breakfast (B)	Chocolate pudding (T, R)	Almonds (NF)
Grapefruit drink (B)	Lemonade (B)	Strawberry drink (B)
Tuna (T)	Ground Beef w/pickle sauce (T)	

S P A C E

U

I

T

S

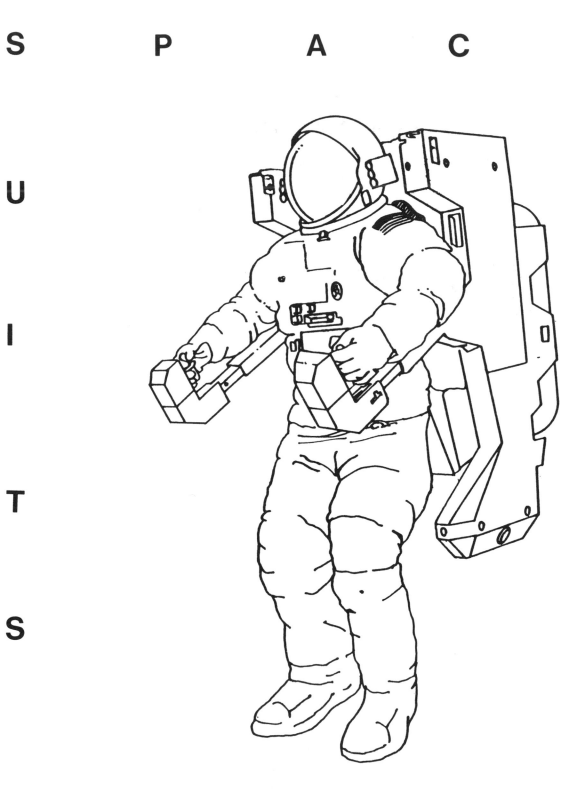

NAME: Space Suits

SKILL: Spelling and Vocabulary

PROCEDURE: Study the parts of the space suit. See how many parts that you can name correctly.

VARIATIONS: Design your own space suit. Don't forget any vital parts.

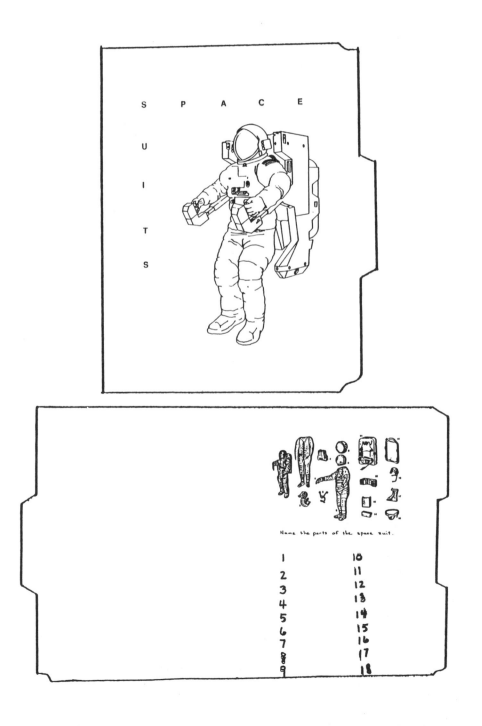

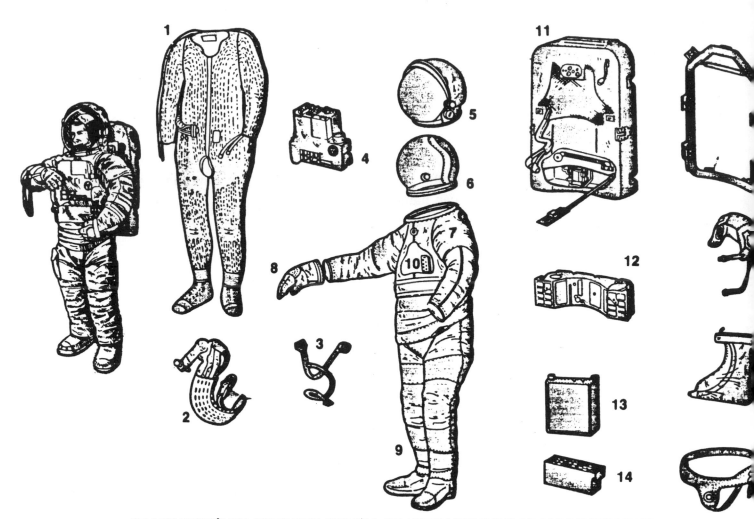

SPACE SUIT/LIFE SUPPORT SYSTEM OR EXTRAVEHICULAR MOBILITY UNIT

1. **Liquid Cooling and Ventilation Garment** worn under the pressure and gas garment. Consists of liquid cooling tubes that maintain desired body temperature.

2. **Service and Cooling Umbilical** contains communications lines; power, water and oxygen recharge lines and a water drain line. It has a multiple connector at one end which attaches to the EMU.

3. **EMU Electrical Harness** provides bioinstrumentation and communications connections to the portable life support system.

4. **Display and Control Module** is a chest mounted control module which contains all external fluid and electrical interfaces, controls and displays.

5. **Extravehicular Visor Assembly** attaches externally to the helmet. Contains visors which are manually adjusted to shield the astronaut's eyes.

6. **Helmet** consists of a clear, polycarbonate bubble, neck disconnect and ventilation pad.

7. **Arm Assembly** contains the shoulder joint and upper arm bearings that permit shoulder mobility as well as the elbow joint and wrist bearing.

8. **Gloves** contain the wrist disconnect, wrist joint and insulation padding for palms and fingers.

9. **Lower Torso** consists of the pants, boots and the hip, knee and ankle joints.

10. **Hard Upper Torso** provides the structural mounting interface for most of the EMU-helmet, arms, lower torso, primary life support subsystem, display and control module, and electrical harness.

11. **Primary Life Support System**, commonly referred to as the "backpack" this assembly contains the life support subsystem expendables and machinery.

12. **Secondary Oxygen Pack** mounted to the base of the primary life support subsystem. It contains a 30-minute emergency oxygen supply and a valve and regulator assembly.

13. **Contaminant Control Cartridge** consists of lithium hydroxide, charcoal, and filters which remove carbon dioxide from the air that the astronaut breathes. It can be replaced in flight.

14. **The Battery** provides all electrical power used by the space suit/life support system. It is filled with electrolyte and charged prior to flight. The battery is rechargeable.

15. **Airlock Adapter Plate** is an EMU storage fixture which is also used as a donning and doffing station.

16. **Communications Carrier Assembly** consists of a microphone and a headset. Allows the astronaut to talk to the other crewmen in the orbiter or other space suit/life support systems.

17. **Insuit Drink Bag** stores liquid in the hard upper torso and has a tube projecting up into the helmet to permit the astronaut to drink while suited.

18. **Urine Collection Device** consists of the adapter tubing, storage bag and disconnect hardware for emptying liquid.

114

ENTERPRISE
COLUMBIA
CHALLENGER
DISCOVERY

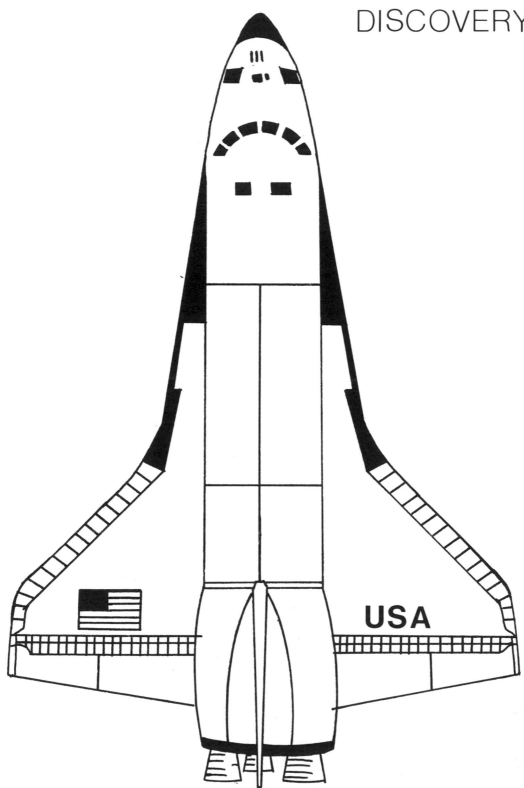

USA

NAME: Space Shuttle

SKILL: Fine Motor Development

PROCEDURE: Glue the shuttle pattern on a file folder or light cardboard. Follow the directions to assemble the shuttle. Fly the model to see what a smooth landing you can make.

VARIATIONS:
1. Read NASA material about the Mercury, Gemini and Apollo space programs which preceded the Space Shuttle.
2. Learn more about the Space Shuttle astronauts from the material received from NASA.
3. Keep a scrapbook on Shuttle flights occuring during the school year. What were some missions of the flight?

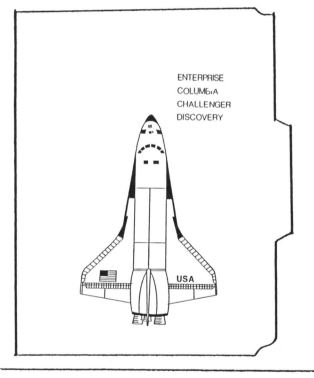

ENTERPRISE
COLUMBIA
CHALLENGER
DISCOVERY

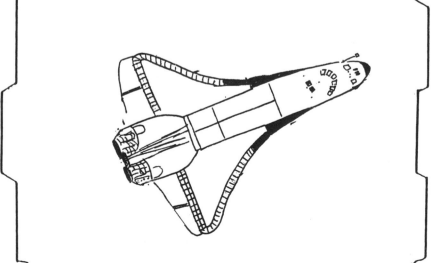

SPACE SHUTTLE MODEL

ASSEMBLY INSTRUCTIONS

Read carefully before assembly:
1. Cut out all parts using scissors.
2. Cut out V-shaped notches on Fuselage to create tabs along the outside edge. Fold tabs out.
3. Glue or tape three Nose Weights to the underside of the nose of your glider. Use the fourth weight provided if needed for extra trim after assembly.
4. Fold Fuselage along middle line.
5. Starting at the nose, glue or tape Fuselage to Deck and Wing Assembly. Match tabs on Fuselage exactly to those printed on Deck and Wing Assembly.
6. To close the nose, glue or tape the two halves together using tabs provided.
7. Fold Vertical Stabilizer Assembly. Fold out tabs A and B. Glue or tape the Vertical Stabilizer assembly to make one solid piece except for tabs A and B.
8. Attach Vertical Stabilizer to Fuselage, matching tab A with point A and tab B with point B.

PREFLIGHT INSTRUCTIONS

For best results, launch your Shuttle glider with a gentle, level toss. Bend the Body Flap up slightly for a greater lift.

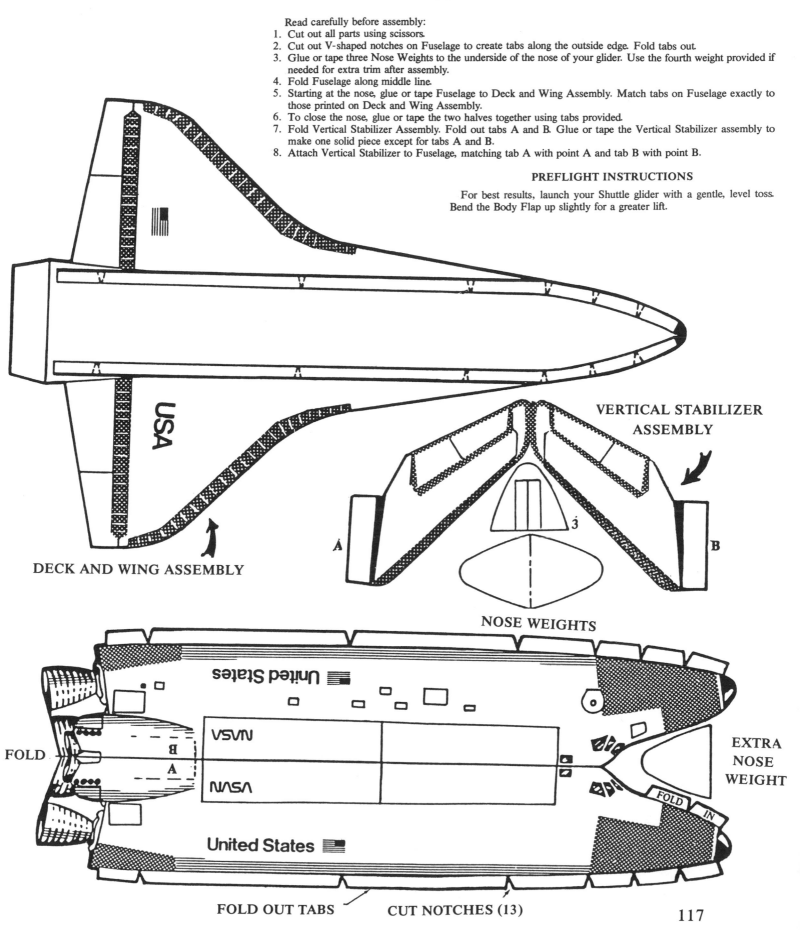

DECK AND WING ASSEMBLY

VERTICAL STABILIZER ASSEMBLY

A
B
3

NOSE WEIGHTS

FOLD

EXTRA NOSE WEIGHT

FOLD OUT TABS   CUT NOTCHES (13)

117

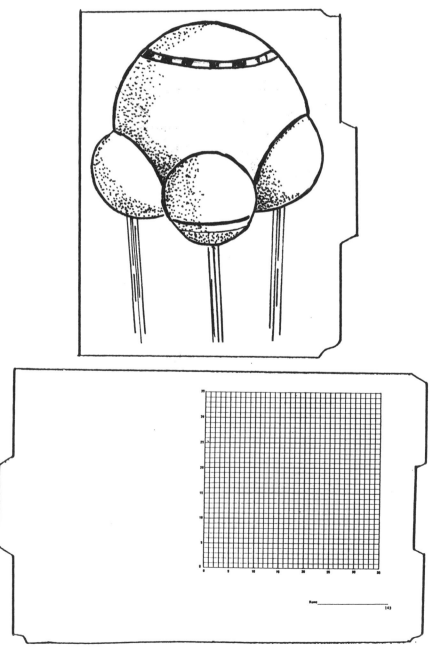

NAME: Space Station

SKILL: Art and Geometry

PROCEDURE: Design your own space station. Be sure to consider adequate space, necessities in space, etc. This sample space station was designed by Emery Leonard when he was in eighth grade.

VARIATIONS: 1. Read Jules Verne's imaginery stories, *Around the World in 80 Days*, *Twenty Thousand Leagues Under the Sea* and *From Earth to the Moon*.
2. Write an imaginary trip into space, stopping at your space station.

UTOPIA

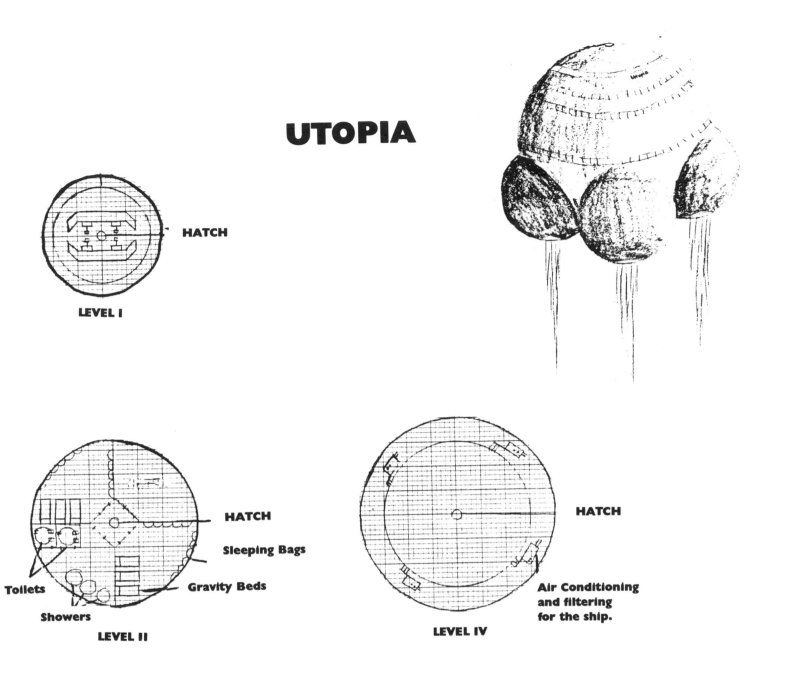

LEVEL I

HATCH

LEVEL II

HATCH

Sleeping Bags

Gravity Beds

Toilets

Showers

LEVEL IV

HATCH

Air Conditioning
and filtering
for the ship.

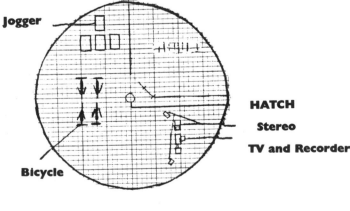

Jogger

HATCH

Stereo

TV and Recorder

Bicycle

LEVEL III

Level 1 – Control Room. There is room for two pilots, a navigator, and a communicator. The floor swivels around for better sight. A watch to the lower levels is in the middle.

Level 2 – Quarters. This ship must hold eleven people. They sleep on the wall for space. There is a male and female room. There are also gravity beds. In case they encounter some high G's they get in these.

Level 3 – Recreation Room. This holds muscle building machines and other various activities for the crew.

Level 4 – Greenhouse. This room provides food and air for the crew.

119

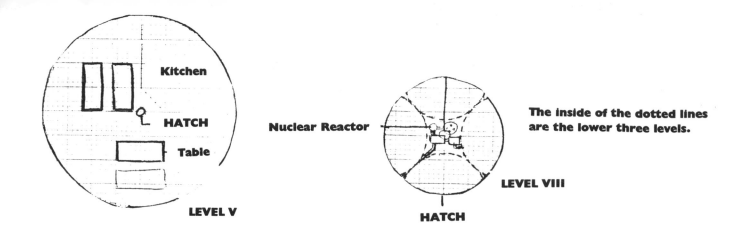

Kitchen

HATCH

Table

LEVEL V

Nuclear Reactor

The inside of the dotted lines are the lower three levels.

LEVEL VIII

HATCH

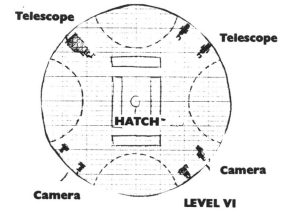

Telescope

Telescope

HATCH

Camera

Camera

LEVEL VI

Level 5 – Cafeteria. This place makes the food and serves it to the crew.
Level 6 – Observatory and Lab. This holds the many experiments to be made and also makes pictures of space and other space bodies.
Level 7 – Shield. This is a solid lead shielding with a lead hatch through it protecting the crew above from harmful radiation.
Level 8 – Powerhouse. Nuclear power is what this level makes converting it into electricity.

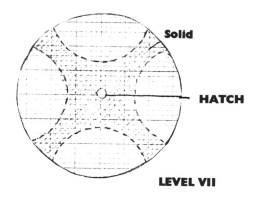

Solid

HATCH

LEVEL VII

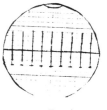

Engines

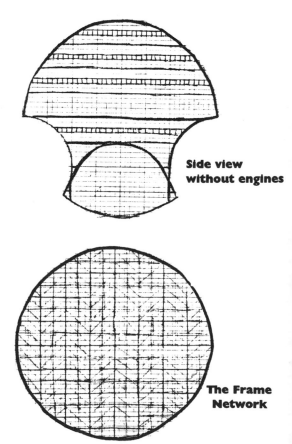

Side view without engines

The frame network is made up of small triangles to permit stability and make it round. The whole craft is shielded with tiles. The walls are thick to insulate and prevent a leak.

The engines work by having a reactor in them and the energy explodes from the explosion of an atom which goes out the hole at the bottom permitting it to move. The control room can control which engine pops thrust and how much there is for steering.

Eleven people are needed to fly the ship. There are two pilots, one navigator, one communicator, two mechanics, two lab technitions, one observationist, and two gardeners. There can also be passengers.

The Frame Network

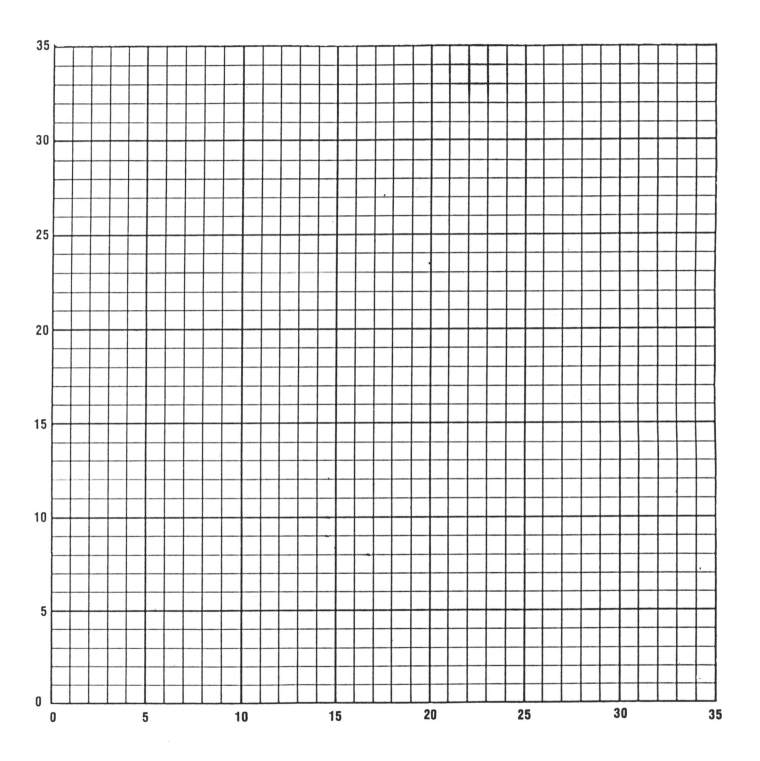

Name_____

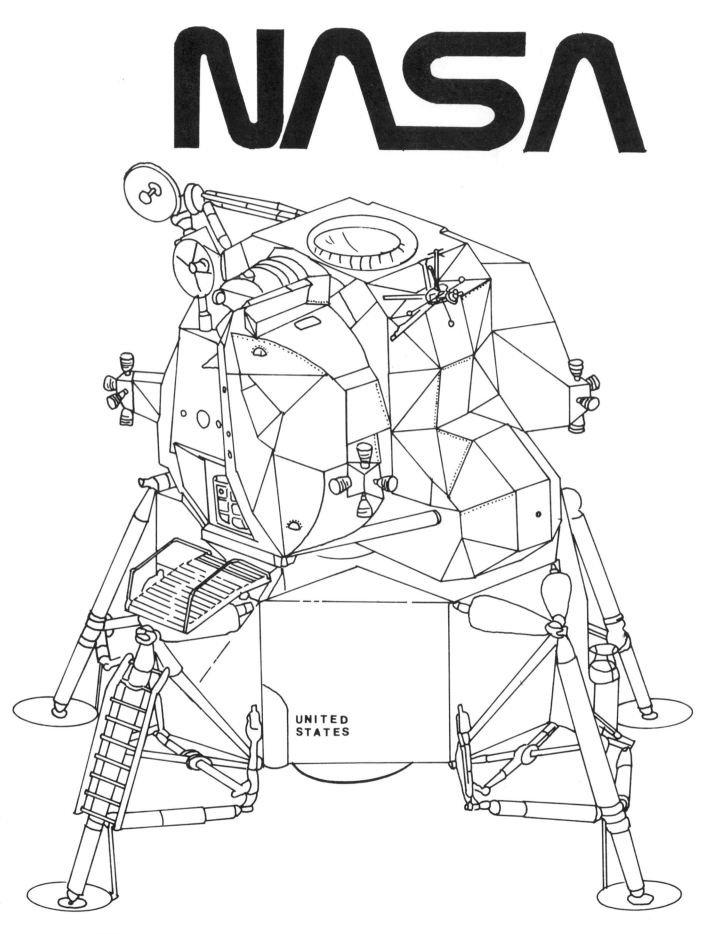

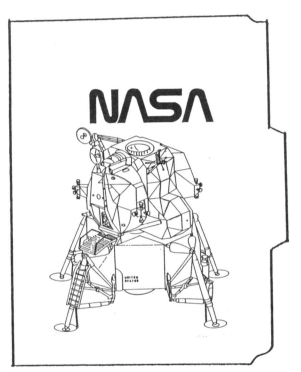

NAME: National Aeronautics and Space Administration (NASA)

SKILL: Awareness of the space program

PROCEDURE: Send the card to NASA to be listed on the teacher's mailing list. Pay special attention to the annual publication, *Spinoffs*, to see the benefits of the space program. List some of these spinoffs such as medical research, solar energy, space food (freeze dried or dried foods).

National Aeronautics and Space Administration
Educational Affairs
Washington, D.C.   20546

NATIONAL AERONAUTICS AND SPACE ADMINISTRATION
REQUEST FOR PLACEMENT ON EDUCATIONAL PUBLICATIONS MAILING LIST

Please place us on your mailing list to receive information about NASA educational publications, films, and other items produced for general distribution.

POSITION TITLE *(Teacher, Principal, Supervisor, Curriculum Director, Librarian, etc.)*

COMPLETE ADDRESS OF SCHOOL OR INSTITUTION <u>WITH CORRECT ZIP CODE</u>

AREAS OF INSTRUCTION *(For teachers only)*	SUBJECT(S)
	GRADE LEVEL(S)

NAME *(If necessary to ensure mail delivery)*   DATE SUBMITTED

NHQ DIV FORM 477

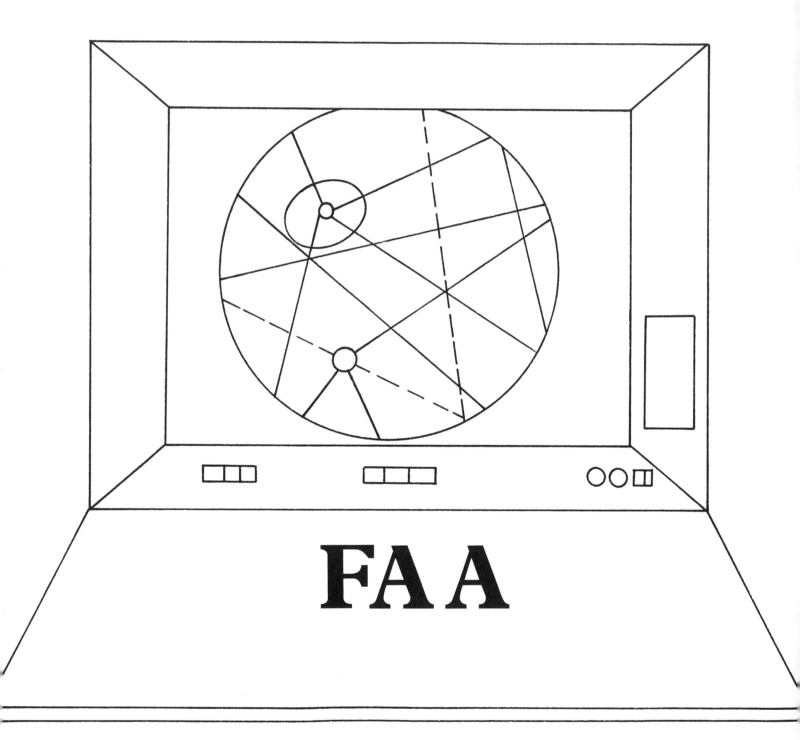

FAA

NAME: Federal Aviation Administration (FAA) Teacher's Guide

SKILL: Integrate aviation into the curriculum: Communication Arts, Science, Social Studies, Health and Career Education.

PROCEDURE: Write the Dep't. of Transportation for free teacher's guide:

FAA
800 Independence Ave., SW
Washington, D.C. 20591

VARIATIONS: Request the order form and request other materials according to your interests and grade level.

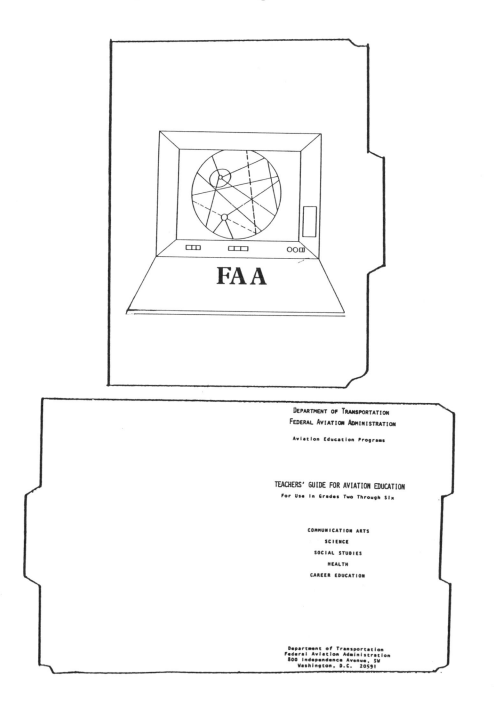

DEPARTMENT OF TRANSPORTATION
FEDERAL AVIATION ADMINISTRATION

Aviation Education Programs

TEACHERS' GUIDE FOR AVIATION EDUCATION
For Use In Grades Two Through Six

COMMUNICATION ARTS
SCIENCE
SOCIAL STUDIES
HEALTH
CAREER EDUCATION

Department of Transportation
Federal Aviation Administration
800 Independence Avenue, SW
Washington, D.C. 20591

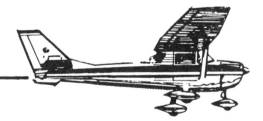

A PROUD TRADITION
OF SERVICE
TO THE NATION

NAME: Civil Air Patrol (CAP) Resources

SKILL: Awareness of the CAP program and resources

PROCEDURE: Order the Falcon Force (Grade 4 – 6) and the coloring book series on aviation giants and other aviation/space concepts.

VARIATIONS: Read about the CAP Cadet Program. Contact your local CAP Squadron by writing Headquarters for contact officer:
HQ   CAP-USAF/PA
Maxwell AFB, AL   36112

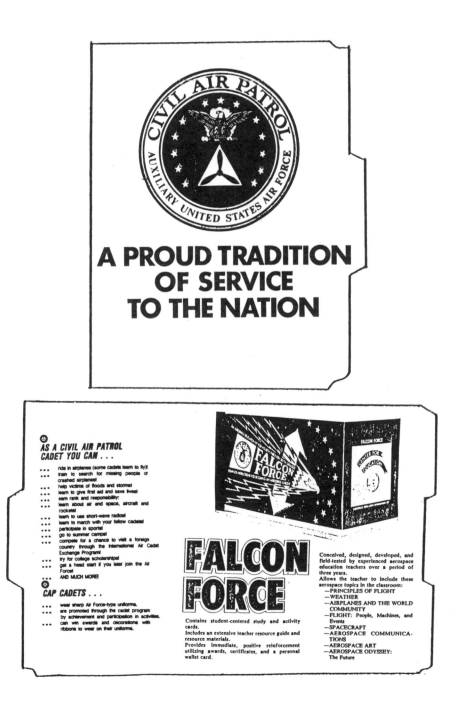

AVIATION

EDUCATION

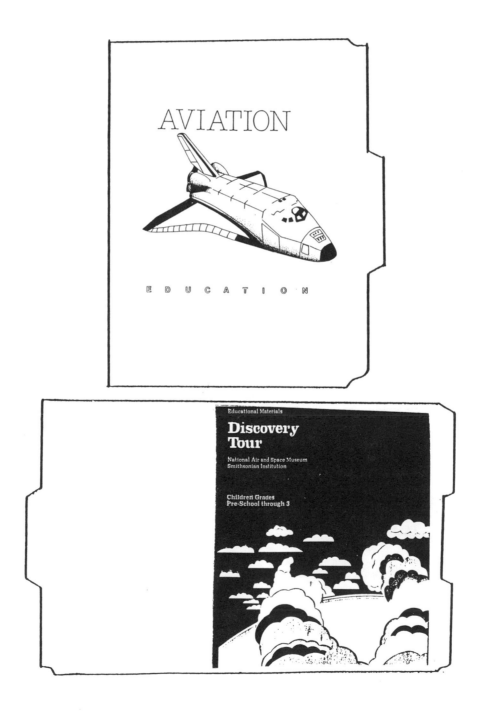

NAME: Smithsonian Air and Space Museum Resources

SKILL: Awareness of Spacecraft

PROCEDURE: Request preschool through grade three, middle grades or high school *Discovery Tour Kits* and other resources.

> Write: National Air and Space Museum
> 7th and Independence Ave., SW
> Washington, D.C.   20560

VARIATIONS: Integrate the numerous activities into the curriculum.

AVIATION

E D U C A T I O N

Educational Materials

Discovery Tour

National Air and Space Museum
Smithsonian Institution

Children Grades
Pre-School through 3

Cessna

NAME: Cessna (Resources)

SKILL: Awareness of flight and related concepts through the *International Air Age Education Packet*.

PROCEDURE: Write for the packets containing the following concepts: How Birds and Aircraft Fly; Parts of the Aircraft; How an Aircraft is Controlled; General Aviation at a Major City Airport; General Aviation at a Small Community Airport; The Total Air Transportation System.

        Cessna Aircraft Co.
        P.O. Box 1521
        Wichita, Kansas   67201

VARIATIONS: Talk to businessmen to see how they use small private airplanes to save time and money for their company.

131

AEROSPACE EDUCATION ASSOCIATION

NAME: Aerospace Education Association

SKILL: Awareness of extensive resources in aerospace education through membership in AEA ($25)

PROCEDURE: Membership includes *Aviation Space Directory* (Annual); *Aviation Space Magazine* (Quarterly); *Aerospace Magazine* (Quarterly); Wings of Aerospace Educator and membership certificate and card.

Write to: AEA
1910 Association Drive
Reston, VA 22091

VARIATIONS: Join the National Space Council.

aerospace projects for the

elementary school

NAME: Aerospace Projects for the Elementary School

SKILL: Interdisciplinary lessons on aviation

PROCEDURE: Read the lesson plan on the various themes: Airline Food, Aviation Language, Researching Aviation, Drama, and Airplane Construction, and incorporate the activities into the classroom. Make the refrigerator box airplane noted in the pictures.

VARIATIONS: Variations to the lesson plan are numerous and listed under activity options.

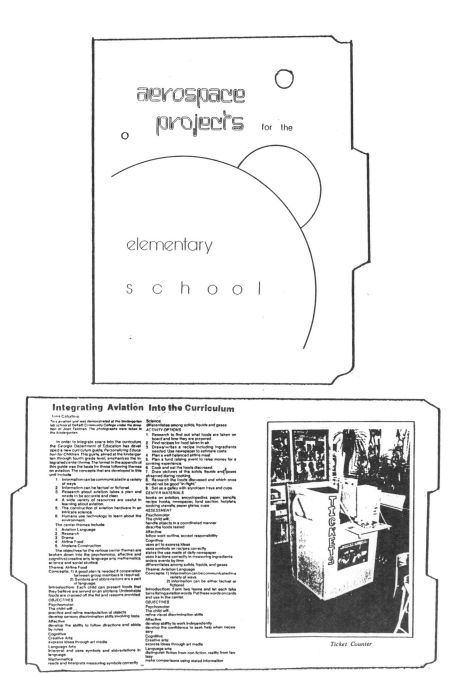

Ticket Counter

Integrating Aviation Into the Curriculum

This aviation unit was demonstrated at the kindergarten lab school at DeKalb Community College under the direction of Jean Feldman. The photographs were taken in the kindergarten.

In order to integrate space into the curriculum, the Georgia Department of Education has developed a new curriculum guide, *Personalizing Education for Children*. This guide, aimed at the kindergarten through fourth grade level, emphasizes the integrated center theme. The format in the appendix of this guide was the basis for these following themes on aviation. The concepts that are developed in this unit include:

1. Information can be communicated in a variety of ways.
2. Information can be factual or fictional.
3. Research about aviation takes a plan and needs to be accurate and clear.
4. A wide variety of resources are useful in learning about aviation.
5. The construction of aviation hardware in an intricate science.
6. Humans use technology to learn about the environment.

The center themes include:

1. Aviation Language
2. Research
3. Drama
4. Airline Food
5. Airplane Construction

The objectives for the various center themes are broken down into the psychomotor, affective and cognitive (creative arts, language arts, mathematics, science and social studies).

Theme: Airline Food

Concepts: 1) A good plan is needed if cooperation between group members is required. 2) Symbols and abbreviations are a part of language.

Introduction: Each child can present foods that they believe are served on an airplane. Undesirable foods are crossed off the list and reasons provided.

OBJECTIVES

Psychomotor:
The child will . . .
- practice and refine manipulation of objects
- develop sensory discrimination skills involving taste

Affective:
- develop the ability to follow directions and abide by rules

Cognitive:

Creative Arts:
- express ideas through art media

Language Arts:
- interpret and uses symbols and abbreviations in language

Mathematics:
- reads and interprets measuring symbols correctly

Science:
- differentiates among solids, liquids and gases

ACTIVITY OPTIONS

1. Research to find out what foods are taken on board and how they are prepared.
2. Find recipes for food taken in air.
3. Draws/writes a recipe including ingredients needed. Use newspaper to estimate costs.
4. Plan a well-balanced airline meal
5. Plan a fund raising event to raise money for a cooking experience.
6. Cook and eat the foods discussed.
7. Draw pictures of the solids, liquids and gases observed during cooking.
8. Research the foods discussed and which ones would not be good "in flight."
9. Set up a galley with styrofoam trays and cups.

CENTER MATERIALS

books on aviation, encyclopedias, paper, pencils, recipe books, newspaper, food section, hotplate, cooking utensils, paper plates, cups

ASSESSMENT

Psychomotor:
The child will . . .
- handle objects in a coordinated manner
- describe foods tasted

Affective:
- follow work outline, accept responsibility

Cognitive:
- uses art to express ideas
- uses symbols on recipes correctly
- states the use made of daily newspaper
- uses fractions correctly in measuring ingredients
- orders events by time
- differentiates among solids, liquids, and gases

Theme: Aviation Language

Concepts: 1) Information can be communicated in a variety of ways.

2) Information can be either factual or fictional.

Introduction: Form two teams and let each take turns listing aviation words. Put these words on cards and use in the center.

OBJECTIVES

Psychomotor:
The child will . . .
- refine visual discrimination skills

Affective:
- develop ability to work independently
- develop the confidence to seek help when necessary

Cognitive:

Creative arts:
- express ideas through art media

Language arts:
- distinguish fiction from non-fiction, reality from fantasy
- make comparisons using stated information

Science:
- extend concepts and vocabulary related to space

ACTIVITY OPTIONS

1. Classify pictures of objects as belonging "on earth" and "in the atmosphere."
2. Read and classify word cards as described above.
3. Use word cards and/or idea starts to write and illustrate a book about air travel.
4. Use word cards to make an illustrated dictionary (pictionary).
5. Draw or paint a factual, aviation picture and write a short story about the picture.
6. Finish a story. i.e., "If I could fly anywhere, I would go to . . " or "Let's take a magic carpet ride. . ."
7. Sort pictures of trains, planes and boats according to type of transportation on land, air and water.
8. Write a description of a trip to another city.
9. Make a filmstrip on air travel in the past, present and future.
10. Brainstorm all the things that fly.

CENTER MATERIALS

task cards, scrap paper, newsprint, tape recorder and tape, books about aviation, word cards, assorted pictures or drawings, crayons, paint, brushes, paper, pencils, story ideas, dictionaries, film and pens for filmstrip

ASSESSMENT

Psychomotor:
The child...
- states which visual cues were used in classifying pictures and/or words

Affective:
- works independently without disturbing others

Cognitive:
- requests help when needed
- differentiates between fact and fiction
- uses at least five aviation related words in tasks completed

Theme: Researching Aviation

Concepts: 1) There are many ways to learn about aviation.
2) Research needs to be accurate.

Introduction: Present aviation books to children and discuss ways that we can find out about air travel. List ideas on a chart or chalkboard.

OBJECTIVES

Psychomotor:
The child will...
- refine handwriting and/or typing skills

Affective:
- develop ability to set realistic goals
- develop ability to use time and resources wisely

Cognitive:

Language arts:
- demonstrate an understanding of various techniques to increase vocabulary

Mathematics:
- place objects in order

Science:
- differentiate among solids, liquids and gases
- demonstrate light control through devices such as mirrors and lenses

Social Studies:
- construct and interpret simple graphs and charts

ACTIVITY OPTIONS

1. Select a topic and develop questions to be answered. Make an outline and share findings.
2. Research and draw an airplane. Label each part.
3. Plan a TV script or mural scene.
4. Research and list the costs of air travel including fuel, materials and labor.
5. Classify pictures/word cards into groups of solid, liquid and gaseous.
6. Visit the playground and describe experiences with gravity and weightlessness (slide, swing, ball toss).
7. Measure a small scale map to determine distance between cities and time required for travel.
8. Make math game by drawing hangers and planes, match numeral on plane with correct hanger, put planes in numerical order.
9. Research various careers: mechanic, flight attendant, office workers, ticket agent, pilot, air traffic controller, etc.
10. Invite a parent in the airline industry to talk to the class.

CENTER MATERIALS

books on aviation, encyclopedias, dictionaries, newsprint, paper, pencils, mural paper, paints, pens, crayons, filmstrips (if available), word cards, chart paper

ASSESSMENT

Psychomotor:

The child . . .

- types or writes legibly.

Affective:

- chooses activities which are on his level.
- completes tasks efficiently.

Cognitive:

- uses more than two sources of information in completing tasks.
- order objects/events by time and distance.
- differentiates among solids, liquids and gases.

Theme: Drama

Concepts: 1) Humans learn to adapt to many environments.

2) Humans learn to adapt to many roles in their environment.

Introduction: Discuss the film "To Fly." Chart the main ideas and write important facts about the film.

OBJECTIVES

Psychomotor:

The child will . . .

- perform basic movements such as locomotor and nonlocomotor skills
- use movement to express creative ideas

Affective:

- display increased confidence in ability to attempt new tasks
- develop the ability to accept praise and/or criticism in a constructive manner.

Cognitive:

Creative arts:

- reacts to various background music with actions and movements
- express ideas through drama

Language arts:

- recognize, recall and retell main ideas, sequence, cause/effect

Social Studies:

- identify, describe and analyze adaptive patterns which emerge as people adapt to the environment

ACTIVITY OPTIONS

1. Read the story about Chuck Yeager, Scott Crossfield, Charles Lindberg and Amelia Earhart. Dramatize these events.
2. Plan and present a dramatization of ideas for future air travel.
3. Listen to records and move to the tempo and rhythm.
4. Describe this music by using fingerpaints or crayons.
5. Create a song using rhythm instruments.
6. Set up an airplane and airport in the drama center. Include: metal detector, passenger seats with seatbelts, magazines, air sickness bags, luggage rack with cardboard suitcases, flight shop, ticket booth with brochures, tickets, baggage tags, and bulletin board with sky scene and travel poster.

CENTER MATERIALS

prop-making materials, paper, pencils, scissors, glue, large boxes, art supplies, records ("I'm Leaving on a Jet Plane," "Up in the Air Junior Birdmen," "Up Like a Rocket," "Captain Entropy"), coloring books on aviation heros (CAP publications)

ASSESSMENT

Psychomotor:
The child . . .
- moves in a coordinated manner
- moves creatively to express ideas

Affective:
- appears confident
- accepts praise and/or criticism

Cognitive:
- reacts to parts of music
- uses drama to express ideas
- portrays main events, details, sequence, and cause/effect relationships in a story

Theme: Airplane Construction
Concepts: 1) Constructing an airplane is an intricate science.
                 2) People use technology in order to explore the environment.
Introduction: Discuss blueprints and what they are used for. Introduce various materials used in construction.

OBJECTIVES
Psychomotor:
The child will . . .
- practice and refine manipulation of objects

Affective:
- develop an understanding of cooperation and planning
- learn to respect others' ideas

Cognitive:
Creative Arts:
- express ideas through art media

Language Arts:
- interpret and use information presented graphically

Mathematics:
- read and make scale drawings

ACTIVITY OPTIONS
1. Draw a blueprint detailing plans for building an airplane.
2. Discover an energy source common to classrooms (rubber band). Experiment to find out how this may be used to thrust object forward.
3. Build a model plane.
4. Research and report on methods of communicating in air.
5. Plan and build a short wave radio and/or telegraph.
6. Make paper airplanes (use Delata Dart Kits) and graph distances achieved by each model.
7. Have a paper airplane competition.

CENTER MATERIALS
blueprint examples, books on planes, rubber bands, lightweight objects for projecting, paper and materials for plane construction, graph paper

ASSESSMENT
Psychomotor:
The child will . . .
- handle objects in a coordinated manner

Affective:
- plan cooperatively
- respect others' ideas

Cognitive:
- uses art to express ideas
- makes diagrams and scale drawings

140 Reference
Personalizing Education for Children, Georgia Department of Education, 1982.

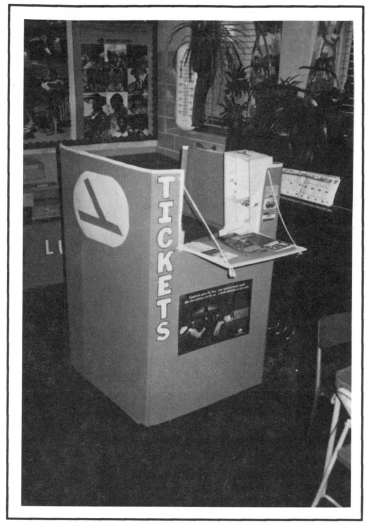

Ticket Counter

Cockpit

141

Luggage Check-in

Galley

NAME: Current Events

SKILL: Reading Comprehension

PROCEDURE: Read the newspaper daily. Cut out aviation/space current events. Ask questions about the article. Write your answers on separate pieces of paper.

VARIATIONS: 1. Have the students ask questions about their own articles.
2. Listen to the news on television and write news articles based on the facts given.

Questions to answer:

How many people have been killed?
1 When did the crash occur?
2 Where did the crash happen?
3 Do authorities know why the crash happened?

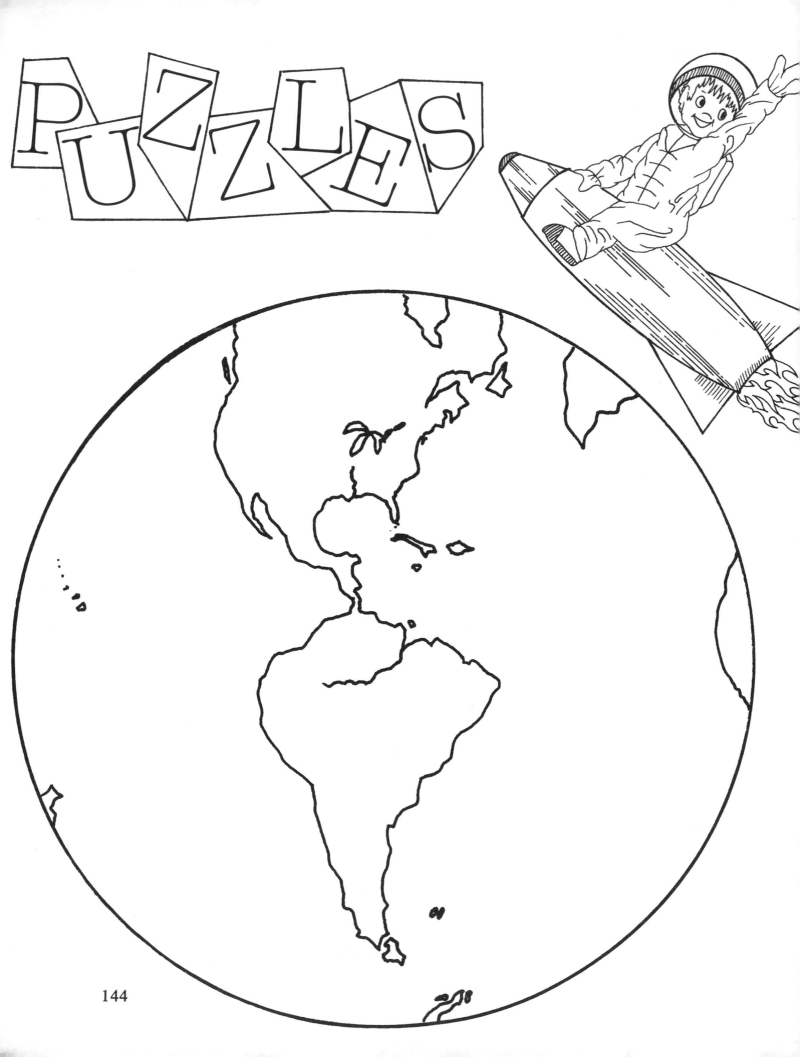

144

NAME: Getting it Together (Puzzles)

SKILL: Fine Motor Coordination

PROCEDURE: Laminate pictures from the resources received. Cut into puzzle pieces.

VARIATIONS: Vary the size and number of puzzle pieces depending on the grade level.

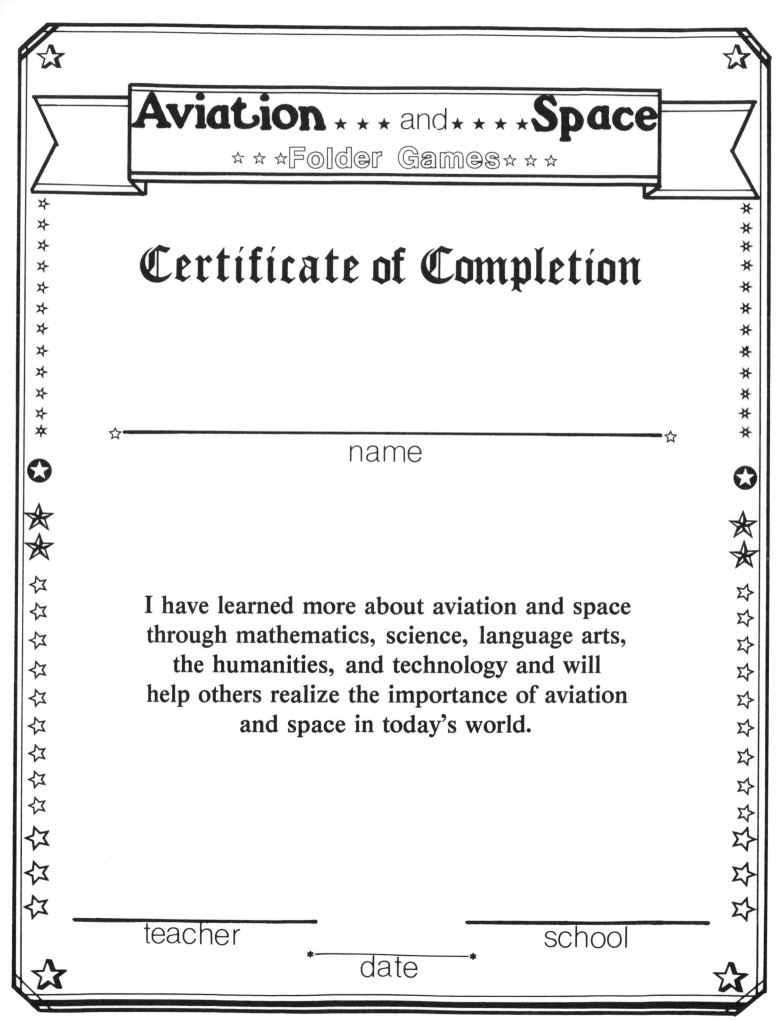

Aviation and Space

Folder Games

Certificate of Completion

_____
name

I have learned more about aviation and space
through mathematics, science, language arts,
the humanities, and technology and will
help others realize the importance of aviation
and space in today's world.

_____   _____
teacher   school

_____
date

VOCABULARY GAMES AND ACTIVITIES

Below is a listing of various vocabulary words relating to aviation and aerospace. You may play various word games with your students in order to help them make these words more familiar to their vocabulary. An asterisk is placed next to the words which may be introduced in the primary grades. Of course, not all children will be able to grasp the meaning of all words and some children will be able to understand words in both lists.

* balloon
* helicopter
* glider
* jet
* rocket
* bi-plane
* airplane
* aircraft
* airlines
* wings
* propeller
  airleron
* elevator
  fuselage
* lights
* engine
* dew
* snow
* rain
* hail
* fog
  air density
* water
* vapor
* degree
* weatherman
  barometer
  knot
  atmosphere
  pressure
  temperature
  humidity

wind velocity
currents
wind
airfoils/maps, charts
forces of flight: weight,
lift, drag, thrust
Bernoulli's Theory
center of gravity
stability
yaw
* airspeed/ground speed
navigation methods: dead
reckoning, pilotage, instruments
airway
approach control
drift
astrodynamics
astrophysics
astronomical
astronautics
aerodynamics
aerobatics
aeronautics
aeroclub
* cumulus
* nimbus
* cirrus
* stratus
nimbo stratus
payload
* orbit
* satellite
hypersonic
weightlessness

* rocket
* planet
* star
* space suit
* astronaut
* blast-off
* countdown
* lift-off
* spacecraft
* take-off
* gravity
* thrust
* zoom
* airport
* aviation
  aerospace
  aviator
* airmen
* runway
* hangar
* ticket
* fuel
* stewards
* pilots
* baggage
  FAA
  CAP
  USAF
  NASA
  FSS
  VFR
  IFR
  VOR

VOCABULARY GAMES AND ACTIVITIES

1. Arrange the words in alphabetical order.
2. Look up the definitions of the words you do not know. Note guide words.
3. Write a sentence with each word.
4. Group the words that pertain to the same subject.
5. Use the words you have grouped in a short story. Use these titles to give you ideas.
   My First Plane Ride
   Airplanes of Yesterday
   Airplanes of Tomorrow
   A Trip to the Moon
   A Trip to Outer Space
   A Trip to Another Planet
   Fighter Planes
   Famous Aviators
   Make up your own title and story.
6. Divide the words into syllables.
7. Tell which words are compound words.
8. Write a question using each word. See if your classmates can answer the question.
9. Use the words you learned to inspire further study. Go to the library and see what you can find.
10. Find out about famous people in aviation and aerospace. Tell about their lives and what they did. Here are a few suggestions: The Wright Brothers, Amelia Earhart, Neil Armstrong, Frank Borman, Goddard, Charles Lindberg, Sally Ride.
11. Study maps, globes, and travel brochures and learn about other states, countries, and planets. Learn about the location, weather, life, and any other interesting facts.

Special note to the teacher:

Many of the stories in this text may be taped and used in the listening center. You may want to make copies of the stories for the children to follow along with as they listen to the tape.

RESOURCES IN AEROSPACE EDUCATION

The Federal Aviation Administration, Office of Public Affairs, Washington, D.C., 20591. Demonstration Aids, First Flight , A Trip to the Airport, Teachers' Guides for Aviation Education, Delta Dart Teacher's Guide, August Martin Activities books, plus many other materials including an extensive career awareness series. (Request order form).

National Aeronautics and Space Administration, Educational Affairs Division, Washington, D.C. 20546. NASA Fact Sheets, Activities for the Elementary Classrooms, Elementary School Aerospace Activities (Lincoln Project), Space Shuttle kits, Spinoffs, plus many other charts, posters and materials. (Request placement on educational publication mailing list).

Civil Air Patrol, National Headquarters, Maxwell Air Force Base, Alabama 36112. Falcon Force, Horizons Unlimited, CAP Newsletter, aviation giants coloring book series, aerospace personality series.

National Air and Space Museum, Smithsonian Institution, Washington, D.C. 20560. Education Materials.

Aerospace Projects for Young Children, Jane Caballero, Humanics Limited, P.O. Box 7447, Atlanta, Georgia 30309.

Aerospace Education Games and Activities for the Elementary School, Pauline Maupin, 2522 Elm Hill Pike, Nashville, Tennessee 37214. ($5.00).

Make 'n Take Experiments, Ted Colton, Aerospace Resource Center, Georgia State University, Atlanta, Georgia 30303.

The Ninety-Nines, An Introduction to Aviation, Inc. P.O. Box 59965, Oklahoma City, Oklahoma 73519. Coloring books (Request various materials).

Delta Dart Project, Midwest Products Co., 400 S. Indiana Street, Hobart, Indiana 46342.

Estes Rocketry and Kite Kits, Estes Industries, Penrose, Colorado 81240.

Jeppesen Sanderson, 56 Inverness Drive East, Englewood, Colorado 80112-5498, for various resources.

Captain Entropy, Bruce Haack, Dimension 5 Records for Children, Box 185 Kingsbridge Station, Bronx, New York 10463. (record).

International Air Age Education Package, Cessna Aircraft Company, P.O. Box 1521, Wichita, Kansas 67201.

Missionspace Teacher's Kit Media Mart, 72 West 45th Street, New York, New York 10036.

Hi-Flyer Manufacturing Company, P.O. Box 280, Penrose, Colorado 81240. (Kite Kits and Clubs).

Humanics Publications

EDUCATION

THE EARLI PROGRAM
Excellent language development program! Volume I contains developmentally sequenced lessons in verbal receptive language; Volume II, expressive language. Use as a primary, supplemental or rehabilitative language program.

Volume I	HL-067-7	$14.95
Volume II	HL-074-X	$14.95

LEARNING ENVIRONMENTS FOR CHILDREN
A practical manual for creating efficient, pleasant and stressfree learning environments for children centers. Make the best possible use of your center's space!

HL-065-0 $12.95

COMPETENCIES:
A Self-Study Guide to Teaching Competencies in Early Childhood Education
This comprehensive guide is ideal for evaluating or improving your competency in early childhood education or preparing for the CDA credential.

HL-024-3 $12.95

LOOKING AT CHILDREN:
Field Experiences in Child Study
A series of fourteen units made up of structured exercises dealing with such issues as language development, play and moral development in children. A fresh new approach to learning materials for early childhood educators.

HL-001-4 $14.95

STORYBOOK CLASSROOMS:
Using Children's Literature in the Learning Center/Primary Grades
A guide to making effective use of children's literature in the classroom. Activities designed for independent use by children K-3, supplemented with illustrations and patterns for easy use. Guidelines, suggestions, and bibliographies will delight and help to instill a love of reading in kids!

HL-043-X $14.95

ACTIVITY BOOKS

EARLY CHILDHOOD ACTIVITIES:
A Treasury of Ideas from Worldwide Sources
A virtual encyclopedia of projects, games and activities for children aged 3-7, containing over 500 different child-tested activities drawn from a variety of teaching systems. The ultimate activity book!

HL-066-9 $16.95

VANILLA MANILA FOLDER GAMES
Make exciting and stimulating **Vanilla Manila Folder Games** quickly and easily with simple manila file folders and colored marking pens. Unique learning activities designed for children aged 3-8.

HL-059-6 $14.95

LEAVES ARE FALLING IN RAINBOWS
Science Activities for Early Childhood
Hundreds of science activities help your children learn concepts and properties of water, air, plants, light, shadows, magnets, sound and electricity. Build on interests when providing science experience and they'll always be eager to learn!

HL-045-6 $14.95

HANDBOOK OF LEARNING ACTIVITIES
Over 125 exciting, enjoyable activities and projects for young children in the areas of math, health and safety, play, movement, science, social studies, art, language development, puppetry and more!

HL-058-8 $14.95

FINGERPLAYS AND RHYMES FOR ALWAYS AND SOMETIMES
More than 250 new fingerplays and rhymes will delight 2 to 5 year olds with rhythm and humor, while teaching shapes, numbers, colors and much more.

HL-083-9 $12.95

ART PROJECTS FOR YOUNG CHILDREN
Build a basic art program of stimulating projects on a limited budget and time schedule with **Art Projects**. Contains over 100 fun-filled projects in the areas of drawing, painting, puppets, clay, printing and more!

HL-051-0 $14.95

BLOOMIN' BULLETIN BOARDS
Stimulate active student participation and learning as you promote your kids' creativity with these delightful and entertaining activities in the areas of Art, Language Arts, Mathematics, Health, Science, Social Studies, and the Holidays. Watch learning skills and self-concepts blossom!

HL-047-2 $12.95

BIRTHDAYS: A CELEBRATION
Stimulate children's creativity and social development with exciting ideas for celebrating birthdays. More than 200 games and activities.

HL-078-2 $12.95

CHILD'S PLAY:
An Activities and Materials Handbook
An eclectic selection of fun-filled activities for preschool children designed to lend excitement to the learning process. Activities include puppets, mobiles, poetry, songs and more.

HL-003-0 $14.95

ENERGY:
A Curriculum for 3, 4 and 5 Year Olds
Help preschool children become aware of what energy is, the sources of energy, the uses of energy and wise energy use with the fun-filled activities, songs and games included in this innovative manual.

HL-069-3 $ 9.95

EXPLORING FEELINGS
Packed with activities that help children acknowledge their feelings and develop self-confidence and positive self-image.

HL-037-5 $14.95

TODDLERS LEARN BY DOING
Hundreds of toddler activities for active play, quiet play, language and concept development and stimulating the five senses.

HL-085-5 $12.95

CHILDREN AROUND THE WORLD
Fun-filled activities introduce children aged 4-7 to cultures and people around the world: their food, clothes, schools, toys and games.

HL-033-2 $14.95

vanilla manila folder games for young children
by Jane A. Caballero

LOOKING AT CHILDREN
field experiences in child study

Storybook Classrooms
Using Children's Literature In the Learning Center
Karla Hawkins Wendelin, Ph.D. • M. Jean Greenlaw, Ph.D.

A Self-study Guide to Teaching Competencies in Early Childhood Education
COMPETENCIES

HUMANICS LIMITED

The Successful Teacher's Most Valuable Resource!

SCISSOR SORCERY
Developmentally-sequenced activities help children (3 – 7) learn to cut. Reproducible activity sheets give children practice cutting lines, shapes, fringes and more.
HL-076-6 $16.95

CAN PIAGET COOK?
More than 40 plans for using simple recipes in the classroom. Children (4 – 8) enjoy science as they measure and observe chemical changes — and eat the results!
HL-078-2 $12.95

-------------------- PARENT INVOLVEMENT

LOVE NOTES
A one year's supply of ready-to-use "Love Notes" to send home with the child to the parent. A novel and charming way to help parents enrich their parenting skills.
HL-068-5 $19.95

PARENTS AND BEGINNING READING
This excellent, highly readable text gives you an overview of children's language development, suggestions for games that enhance reading skills, ideas for establishing a reading environment in the home, tips for grandparents, and lists of resources.
HL-044-8 $14.95

WORKING TOGETHER:
A Guide to Parent Involvement
Ideal guide for those wishing to launch a new parent involvement program or improve existing parent/ school communication and interaction. Favorably reviewed by the National Association for Young Children.
HL-002-2 $14.95

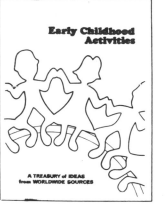

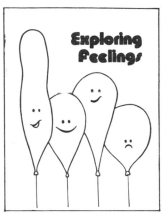

-------------------- ASSESSMENT

THE LOLLIPOP TEST
A Diagnostic Screening Test of School Readiness
Based on the latest research in school readiness, this test effectively measures children's readiness strengths and weaknesses. Included is all you need to give, score and interpret the test.
HL-028-6 $19.95

THE PRESCHOOL ASSESSMENT HANDBOOK
Combines vital child development concepts into one integrated system of child observation and assessment. This is also the user's guide to the Humanics National Child Assessment Form— Ages 3 to 6.
HL-039-1 $16.95

THE INFANT/ TODDLER ASSESSMENT HANDBOOK
User's guide to the Humanics National Child Assessment Form— Ages 0 to 3. Integrates crucial concepts of child development into one effective system of observation and assessment.
HL-049-9 $14.95

-------------------- SOCIAL SERVICES

HUMANICS LIMITED SYSTEM FOR RECORD KEEPING
Designed to meet all record-keeping needs of family-oriented social service agencies, this guide integrates the child, family, social worker and community into one coherent network. Also the user's guide to proper use of Humanics Limited Record Keeping Forms.
HL-027-8 $12.95

REAL TALK
Exercises in Friendship and Helping Skills
Real Talk teaches students basic skills in interpersonal relationships though such methods as role-playing and modeling. An ideal human relations course for elementary, junior high and high schools.

| Teacher's Manual | HL-026-X | $ 7.95 |
| Student's Manual | HL-025-1 | $12.95 |